ANNALS *of* THE NEW YORK ACADEMY OF SCIENCES

T0127744

EDITOR-IN-CHIEF
Douglas Braaten

ASSOCIATE EDITOR
Rebecca E. Cooney

PROJECT MANAGER
Steven E. Bohall

Artwork and design by Ash Ayman Shairzay

The New York Academy of Sciences
7 World Trade Center
250 Greenwich Street, 40th Floor
New York, NY 10007-2157

annals@nyas.org
www.nyas.org/annals

**The New York
Academy of Sciences**

Published by Blackwell Publishing
On behalf of the New York Academy of Sciences

Boston, Massachusetts
2012

ANNALS *of* THE NEW YORK ACADEMY OF SCIENCES

VOLUME
1269

ISSUE

Thymosins in Health and Disease I

The Third International Symposium

ISSUE EDITORS

Allan L. Goldstein[a] and Enrico Garaci[b]

[a]The George Washington University School of Medicine and Health Sciences
[b]Istituto Superiore di Sanità

TABLE OF CONTENTS

Academy Membership: Connecting you to the nexus of scientific innovation

Since 1817, the Academy has carried out its mission to bring together extraordinary people working at the frontiers of discovery. Members gain recognition by joining a thriving community of over 25,000 scientists. Academy members also access unique member benefits.

 Network and exchange ideas with the leaders of academia and industry

 Broaden your knowledge across many disciplines

 Gain access to exclusive online content

 Select one free *Annals* volume each year of membership and get additional volumes for just $25

Join or renew today at **www.nyas.org**.
Or by phone at **800.843.6927** (**212.298.8640** if outside the US).

Ann. N.Y. Acad. Sci. ISSN 0077-8923

ANNALS OF THE NEW YORK ACADEMY OF SCIENCES
Issue: *Thymosins in Health and Disease*

Introduction for *Thymosins in Health and Disease*

The Third International Symposium on Thymosins in Health and Disease brought together many of the leading scientists, clinicians, and thought-leaders from North and South America, Europe, and Asia to discuss the latest advances and clinical applications of the thymosins in both basic and clinical areas. The symposium, held in Washington, DC on March 14–16, 2012, was sponsored by the George Washington University, the Istituto Superiore di Sanitá (ISS), and the University of Rome "Tor Vergata."

Of the six plenary sessions, three were devoted to advances in basic research and two were devoted to translational advances in the broad areas of immunization, immune deficiency diseases, infectious diseases, cancer, wound healing, and treatment of eye injuries, cardiovascular diseases, and neurological injuries. A sixth, late-breaking, basic research session focused on advances in nuclear magnetic resonance analysis of thymosin peptides, novel antimicrobial agents, use of stem cells and growth factors for tissue and burn repair, and advances in analysis of peptides.

Luigina Romani, professor of microbiology at the University of Perugia, and Deepak Srivastava, director of the Gladstone Institute of Cardiovascular Disease at the University of California, San Francisco, gave keynote lectures and were co-recipients of the 2012 Abraham White Distinguished Science Award, which was presented at the Awards Banquet. The 2012 Abraham White Distinguished Humanitarian Award was given posthumously to Michael Stern and was accepted by his daughter Margaret Stern. Mr. Stern was recognized for his lifetime of good deeds aimed at supporting and improving Italian–American scientific relations and for his journalistic excellence and philanthropic endeavors. Stern played an important role in co-creating the Fisher Center for Alzheimer's Research Foundation at the Rockefeller University and in starting the Michael Stern Parkinson's Research Foundation.

Dr. Romani was honored for her pioneering studies and scientific contributions, which have significantly increased our understanding of the molecular mechanism of thymosin $\alpha 1$ (Tα1), and for elucidating the important role it plays in modulating Toll-like receptors and in immunity. Her breakthrough discovery of tryptophan metabolites as potential antifungal strategies, and her most recent report of functional genomics to predict risk for fungal infections in transplant patients have provided the biomedical community with new strategies for the treatment of serious fungal diseases.

Dr. Srivastava was honored for his pioneering studies and scientific contributions, which have significantly advanced our understanding of the role of thymosin $\beta 4$ (Tβ4) in the development and functioning of the human heart. His studies have provided a novel therapeutic target in the setting of acute myocardial damage. His most recent studies on the molecular events regulating early and late developmental decisions that instruct progenitor cells to adopt a cardiac cell fate and, subsequently, to fashion a functioning heart hold much potential in preventing congenital defects and treating acquired heart disease, particularly with cardiac-specific differentiation of embryonic stem cells. Most recently, he has been able to reprogram adult fibroblasts directly into cardiomyocyte-like cells for regenerative purposes.

doi: 10.1111/j.1749-6632.2012.06729.x
Ann. N.Y. Acad. Sci. 1269 (2012) vii-ix © 2012 New York Academy of Sciences.

The conference scientific summary

The papers included in the two volumes of these proceedings highlight the progress that has occurred in the biological, chemical, and clinical applications of the thymosins in health and disease. Volume 1269 (plenary sessions I, II, and III) includes papers that summarize the most recent advances in basic research, which have provided the scientific foundation for translational studies with several of the thymosin peptides. Volume 1270 (plenary sessions IV, V, and VI) includes papers on clinical applications of thymosins $\alpha 1$ and thymosin $\beta 4$. This volume also has a number of reports describing novel labeling techniques, formulations, and diagnostics, as well as the characterization of a thymosin $\beta 4$ fragment with antimicrobial properties isolated from the sea urchin *Paracentrotus lividus*.

Plenary session I focused on advances in our understanding of the family history of the β-thymosins and the chemistry and role of prothymosin α (ProTα), T$\alpha 1$, and T$\beta 4$ on cellular receptors, signaling, and multifunctionality. John Edwards gave a brilliant lecture on the origin of the β-thymosin family, tracing its evolution from single-celled organisms to fish and land vertebrates. Milica Radisic reported on a significant advance in the engineering and encapsulation of T$\beta 4$ in a collagen–chitosan hydrogel, which allows for controlled release of T$\beta 4$ at nearly zero order kinetics. The presentations by Francesca Fallarino and Enrico Garaci focused on the interface between T$\alpha 1$ and Toll-like receptors, and the role of T$\alpha 1$ in enhancing antitumor activity. In his presentation, Yeu Su focused on upregulation of T$\beta 4$ in a colon cancer model and the use of T$\beta 4$ as a potential therapeutic marker for human colon cancers. In the last plenary talk of the session, Hiroshi Ueda presented new evidence that ProTα is a multifunctional nuclear protein that has important activities in the central nervous system.

Plenary session II focused on advances in our understanding of key molecular markers and the activity of T$\beta 4$ in accelerating wound healing, reducing fibrosis, and inflammation in injured tissues. Gavino Faa began the session with a talk covering the general expression of T$\beta 4$ in human tissues. Karina Reyes-Gordillo discussed T$\beta 4$ and the mechanism by which it prevents liver fibrosis in a rat model using carbon tetrachloride. Enrico Conte showed us through his *in vivo* studies that, in the lung, T$\beta 4$ protects from bleomycin-induced damage in C57BL/6 mice. Sprague Hazard's presentation provided new evidence that the myofibroblast plays an important role in scar formation and the mechanism by which T$\beta 4$ reduces scar formation following injury. In the final presentation of the session, Hee-Jae Cha discussed the stabilization of the hypoxia-inducible factor-1α by T$\beta 4$ through an oxygen-independent manner.

Plenary session III focused on the activities of T$\beta 4$ in cardiovascular protection, neuroplasticity, regeneration, and stem cell differentiation. Sudhiranjan Gupta began the session with a very informative talk on how T$\beta 4$ protects the heart and reduces inflammation and fibrosis. Nicola Smart then discussed the facilitation of myocardial regeneration by T$\beta 4$ via activated epicardial progenitor cells. Christoffer Stark presented data on the cardioprotective effects of T$\beta 4$ after myocardial infarction. Christian Kupatt-Jeremais presented new data documenting the essential elements of T$\beta 4$-mediated collateral growth. In studies of the brain following ischemic injury, Dan Morris outlined the mechanism by which T$\beta 4$ mediated oligodendrogenesis after stroke. David Bader closed the session with a fascinating presentation showing how T$\beta 4$ activates mesothelial cells from the omentum to repair damaged blood vessels.

Plenary session IV focused on T$\alpha 1$ in areas of immunomodulation, immunopharmacology, infectious diseases, and cancer. Virgil Dalm gave the first talk of the session on the gene expression abnormalities of motility disturbed monocytes in recurrent and chronic purulent rhinosinusitis and its correction by thymosin fraction 5 and T$\alpha 1$. Michele Maio then discussed the role of T$\alpha 1$ in treating patients with advanced melanoma. Annalucia Serafino's presentation outlined studies documenting the role of T$\alpha 1$ in innate cellular immunity. Luigina Romani, one of the keynote speakers, reported on the first clinical trial of bone marrow–transplanted patients treated with T$\alpha 1$

to prevent infections. Cynthia Tuthill provided an excellent summary of the pleotropic activities of Tα1 and the activity of Tα1 as an enhancer of vaccine responses. Roberto Camerini ended the session with a presentation of the results of a late stage clinical trial of Tα1 as an adjuvant for enhancing the efficacy of the H1N1 vaccine in patients with end-stage renal disease.

Plenary session V dealt with the clinical and preclinical applications of Tβ4 and its role in wound healing, repair of eye injuries, and Tβ4's effects in treating cardiovascular disease and neurological injuries. Mark Hardy started the session by presenting the results of a study of combination therapy with Tβ4 in an animal model. He provided the first experimental evidence indicating that using Tβ4 in combination with silver sulfadiazine enhances the healing of acute and infected wounds. Terry Treadwell discussed the results of two phase II trials providing the first clinical evidence that Tβ4 accelerates the rate of healing of both pressure and venous stasis ulcers. Gabe Sosne focused his remarks on Tβ4 as a potential novel dry eye therapy. Ye Xiong discussed Tβ4's improvement of functional recovery after traumatic brain injury, as well as its neuroprotective and neurorestorative effects. Daniel Stromberg concluded the session with his presentation on the proposed use of Tβ4 as a cardioprotectant during congenital heart surgery in a randomized, double-blind clinical trial.

Plenary session VI was a late-breaking paper session including such topics as the NMR structure of human Tα1, presented by David Volk; the isolation of a novel β-thymosin from the sea urchin *Paracentrotus lividus* and its use as a novel antimicrobial peptide against *Staphylococcal* biofilms, presented by Domenico Schillaci; and Tβ4's key role in the regulation of actin polymerization and depolymerization, and the development of an analytical HPLC methodology to study actin in the sputum of cystic fibrosis patients, presented by Ali Damavandy. In additional presentations, Javier Paino presented preliminary data on the re-epithelialization of burn injuries using cultured umbilical cord cells; Christine App and Jana Knop presented data on Tβ4 labeling and molecular investigations using Tβ4 antibodies. Ending the session was a talk by Jeffrey Thatcher about implantable poly(lactide-co-glycolide) microspheres for sustained release of Tβ4.

The increasing number of thymosin peptides available synthetically has significantly accelerated animal model experimentation in the field and has helped researchers to consider novel clinical applications. These *Annals* volumes should be of keen interest to basic and clinical scientists, pharmacologists, immunologists, biochemists, and cell biologists with interests in the area of biological response modifiers. It should also be of special interest to physicians studying the clinical applications of the thymosins in both health and disease.

We would like to thank the program committee: Roberto Camerini, David Crockford, Paul Riley, Ewald Hannappel, Hynda Kleinman, Luigina Romani, Cynthia Tuthill, Deepak Srivastava, and Bruce Zetter for their help in planning this important scientific symposium. We would also like to acknowledge the editorial staff of *Annals of the New York Academy of Sciences* for their support in editing and publishing this issue.

We are very appreciative of the generous support of the following organizations: the George Washington University; the University of Rome "Tor Vergata;" University of Catania; the Istituto Superiore di Sanitá; Regina Elena Cancer Hospital; Sigma-Tau Pharmaceuticals, Inc.; SciClone Pharmaceuticals, Inc.; RegeneRx Biopharmaceuticals, Inc.; Rothwell, Figg, Ernst & Manbeck, P.C.; Fisher Clinical Services, Inc.; and PolyPeptide Laboratories, Inc. Without the support of our friends from the private sector, this type of educational meeting would not be possible.

ALLAN L. GOLDSTEIN
The George Washington University, Washington, DC

ENRICO GARACI
Istituto Superiore di Sanitá, Rome, Italy

Ann. N.Y. Acad. Sci. ISSN 0077-8923

ANNALS OF THE NEW YORK ACADEMY OF SCIENCES

Issue: *Thymosins in Health and Disease*

Jack of all trades: thymosin α1 and its pleiotropy

Luigina Romani,[1] Silvia Moretti,[1] Francesca Fallarino,[1] Silvia Bozza,[1] Loredana Ruggeri,[2] Andrea Casagrande,[1,3] Franco Aversa,[2] Francesco Bistoni,[1] Andrea Velardi,[2] and Enrico Garaci[4]

[1]Department of Experimental Medicine and Biochemical Science, University of Perugia, Perugia, Italy. [2]Department of Clinical and Experimental Medicine, University of Perugia, Perugia, Italy. [3]Istituto Superiore di Sanità, Roma, Italy. [4]Department of Experimental Medicine and Biochemical Science, University of Tor Vergata, Rome, Italy

Address for correspondence: Luigina Romani, M.D., Ph.D., Department of Experimental Medicine and Biochemical Sciences, University of Perugia, Via del Giochetto, 06122 Perugia, Italy. lromani@unipg.it

Thymosin α1 (Tα1), a thymosin-related 28-mer synthetic amino-terminal acetylated peptide, has gained increasing interest in recent years, due to its pleiotropy. The peptide has been used worldwide as an adjuvant or immunotherapeutic agent to treat disparate human diseases, including viral infections, immunodeficiencies, and malignancies. The peptide can enhance T cell, dendritic cell (DC), and antibody responses, modulate cytokine and chemokine production, and block steroid-induced apoptosis of thymocytes. Its central role in modulating DC function and activating multiple signaling pathways that contribute to different functions may offer a plausible explanation for its pleiotropic action. Additionally, the ability of Tα1 to activate the indoleamine 2,3-dioxygenase enzyme—which confers immune tolerance during transplantation and restrains the vicious circle of chronic inflammation—has been a turning point, suggesting a potential, specific function in immunity. Accordingly, Tα1 has recently been shown to promote immune reconstitution and improve survival of recipients of HLA-matched sibling T cell–depleted stem cell transplants in a phase I/II clinical trial. Thus, Tα1 continues to live up to its promises.

Keywords: thymosin α1; hematopoietic transplantation; IDO; TLRs; tolerance

Introduction

In the early 1960s, the thymus came of age with pioneering work on its importance in the development of the lymphoid system. At the same time, seminal work by Goldstein *et al.* led to the identification and characterization of a thymic lymphopoietic factor (thymosin α1), whose characterization was later expanded upon by Garaci's group.[1,2] Pleiotropy (from the Greek πλείων, meaning more, and τροπή, meaning turn or convert) is the property of a protein to display various and sometimes opposing effects. In this regard, Tα1, as a 28-mer synthetic amino-terminal acetylated peptide, has been used worldwide as an adjuvant or immunotherapeutic agent to treat disparate human diseases, including viral infections, immunodeficiencies, and malignancies. Tα1 has shown an excellent safety profile and does not induce the side effects and toxicities commonly associated with immunomodulatory agents.[3] Indeed, the advent of human recombinant immune modulators held promise, but the outcomes of clinical trials using biologics that target individual immune mediators have been disappointing. The complex pathophysiology of inflammatory-driven immune and autoimmune diseases, from infections to cancer, is self-amplifying and redundant at multiple levels. Thus, the intrinsic robustness of living systems against various perturbations is a key factor that prevents such compounds from being successful. This implicates that pharmacologic therapy for inflammatory diseases would only be successful if addressing these complex network systems. Any successful drug would have to be pleiotropic, and work on the many components of the inflammatory/autoimmune cascade at a time. Tα1 exceptionally fulfills this requirement, qualifying itself as a context-dependent molecule capable of

doi: 10.1111/j.1749-6632.2012.06716.x

multitargeted interactions in physiological and non-physiological conditions. In this review, we discuss new insights into the molecular mechanisms of Tα1's pleiotropy.

The multifaceted interface between Tα1, dendritic cells, and Toll-like receptors

The mechanism of action of Tα1 is thought to be related to its immunomodulating activities, centered primarily on the augmentation of T cell function (Fig. 1). However, mechanistically, Tα1 has shown an action beyond its effect on T lymphocytes to include an ability to act as an endogenous regulator of both the innate and adaptive immune systems (Fig. 2).[4,5] DCs have a crucial role in determining immune outcomes by acquiring antigens, collating environmental cues, and then becoming cells that are either potent stimulators or suppressors of T cell responses.[6] Tα1 has shown the ability to modulate signals delivered through innate immune receptors on DCs, thus affecting both innate and adaptive immune responses.

One of the most basic mechanisms for activation of the DC system is through the Toll-like receptors (TLRs). TLRs belong to the type I transmembrane receptor family. Their expression is ubiquitous, from epithelial to immune cells. The TLR family members are pattern recognition receptors that collectively recognize lipid, carbohydrate, peptide, and nucleic acid structures that are broadly expressed by different groups of microorganisms. Some TLRs are expressed at the cell surface, whereas others are expressed on the membrane of endocytic vesicles or other intracellular organelles. TLR triggering induces DC maturation, which leads to the upregulation of costimulatory molecules such as CD40, CD80, and CD86, and secretion of immune modulatory cytokines and chemokines. Two major signaling pathways are generally activated in response to TLR ligands.[7] One pathway involves the myeloid differentiation factor 88 (MyD88) pathway to activate nuclear factor-kappa B (NF-κB), JUN kinase (JNK), and p38, finally resulting in the production of pro-inflammatory cytokines such as tumor necrosis factor (TNF)-α, interleukin (IL)-12, and IL-1, and induction of innate effector mechanisms. The other uses the MyD88-independent production of type I interferons via the TIR-domain–containing adapter-inducing interferon-β (TRIF) pathway. In addition, TLRs can directly stimulate the proliferation of CD4[+] and CD8[+] T cells as well as reverse the suppressive function of T$_{reg}$ cells.[8]

Given the ability of Tα1 to activate TLRs, the direct responsiveness of T lymphocytes to TLR ligands

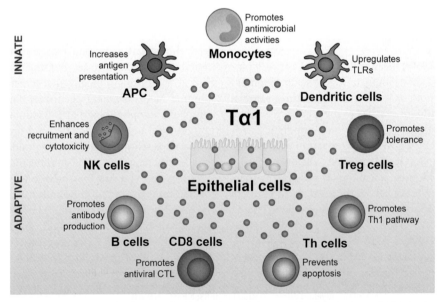

Figure 1. Pleiotropic immune activation by Tα. The drug actions of Tα1 on cells of the innate and adaptive immune system. This is a pictorial representation of all literature-supported actions of Tα1 on immune cells. APC, antigen presenting cells; CTL, cytotoxic T lymphocytes; NK cells, natural killer cells; Th cells, T helper cells; T$_{reg}$ cells, regulatory T cells; TLRs, Toll-like receptors.

may offer new perspectives for the immunothera-peutic manipulation of T cell responses by Tα1. Tα1 strongly upregulated the expression of *TLR2* and *TLR9* by murine and human DC subsets and activated NF-κB and JNK/p38/AP-1 pathways.[9,10] Induction of *TLR2* or *TLR9* expression by Tα1 was associated with a distinct activation program in DC subsets and involved both the MyD88 and TRIF pathways. Tα1 induced the production of IL-12p70 and IL-10 in a TLR-dependent manner. Production of IL-12p70 was reduced in *TLR2*[−/−] mice, whereas IL-10 was reduced in *TLR9*[−/−] mice. Our preliminary findings seem to suggest that the abil-ity of Tα1 to signal through disparate TLRs on DC subsets may occur through the interaction with multiple, yet distinct, TLR ligands, either endoge-nous or exogenous (Fig. 2). This finding not only qualifies Tα1 as a context-oriented molecule but may also offer molecular explanations for a success-ful drug working on the many components of the inflammatory/immune response, that is, for suc-cessful pleiotropy.

The immunomodulatory effects of Tα1 on DCs correlated with a therapeutic effect of the peptide in experimental fungal or viral infections.[5] Admin-istration of Tα1 to mice with *Aspergillus fumigatus*

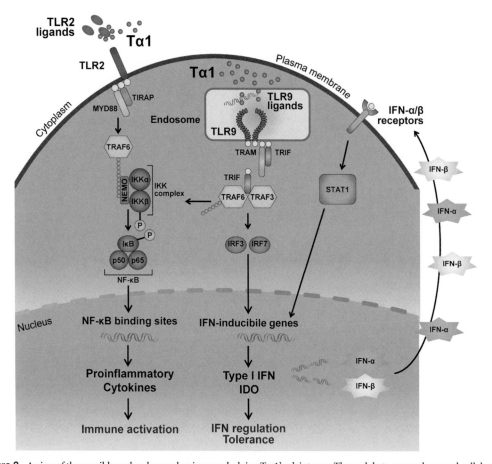

Figure 2. A view of the possible molecular mechanisms underlying Tα1's pleiotropy. Through heterocomplexes and collaborative interactions between endogenous and/or exogenous TLR agonists, Tα1 may lead to the activation of different members of the NF-κB and IRF families that plays a central role in the stress response and inflammation and transcription of interferons. Thus, the intersection between canonical and noncanonical signaling pathways may be crucial in promoting an optimally protective response balanced between inflammation and tolerance by Tα1. IDO, indoleamine 2,3-dioxygenase; IKK, Inhibitory kynase; IRF, Interferon regulatory factor; MyD88, myeloid differentiation factor 88; NEMO, NF-κB essential modulator; NF-κB, nuclear factor-κB; TRIF, TIR-domain-containing adapter-inducing interferon-β; STAT1, Signal Transducer and Activator of Transcription; TRAF, TNF receptor associated factors.; TRAM, TRIF-related adaptor molecule; TIRAP, Toll-interleukin 1 receptor domain containing adaptor protein.

resulted in a state of full protection to the fungus as revealed by the increased survival after the infection that paralleled the reduced fungal growth and the promotion of interferon (IFN)-γ–producing Th1 cells.[9] The achievement of full protection to the fungus required the coordinated action of several TLRs on innate effector cells. Consistent with the *in vitro* data on TLR expression, the therapeutic efficacy of Tα1 *in vivo* was dependent on MyD88 signaling activated by TLR2 and partially TLR9.[9] Therefore, despite a degree of redundancy in the TLR usage, the MyD88-dependent signaling pathway is essential in the antifungal activity of Tα1 *in vitro* and *in vivo*. In contrast, the ability of Tα1 to modulate DC functioning through TLR9 essentially correlated with an effect on cytomegalovirus (CMV) sensing by DCs *in vitro* and *in vivo*.[11] The antiviral effect of Tα1 led to the activation of interferon regulatory factor (IRF)7 and the promotion of the IFN-α/IFN-γ–dependent effector pathway in plasmacytoid DCs via the TLR9-dependent viral recognition sensing. After infection, Tα1 decreased the viral load in mice with primary murine CMV infection and induced the expansion of cytolytic NK1.1$^+$ cells, IFN-γ–producing CD8$^+$ or CD4$^+$ T cells, and the production of IL-12p70, IFN-α, IFN-γ, and IL-10.[11] Together, these data indicate that Tα1 affects both the innate and adaptive antimicrobial immune responses *in vivo*.

En route to tolerance: Tα1 meets IDO and induces T$_{reg}$ cells

The induction of tolerance is critical for the maintenance of immune homeostasis. DCs not only play a key role in the induction of immune responses, but also in the induction and maintenance of immune tolerance.[6] Recent studies have demonstrated a crucial role for tryptophan catabolism and kynurenine production in the induction of peripheral tolerance.[12,13] The enzyme indoleamine 2,3-dioxygenase (IDO) is an intracellular heme-containing enzyme that catalyzes the initial and rate-limiting step in tryptophan degradation along the kynurenine pathway. IDO is widely expressed in a variety of human tissues as well as in macrophages and DCs, and is induced in inflammatory states by IFN-γ and other proinflammatory cytokines. Work has demonstrated a complex and crucial role for IDO in immunoregulation during infection, pregnancy, autoimmunity, transplantation, and neopla-

sia.[13] TLR9 stimulation may lead to the activation of the tryptophan catabolism pathway in DCs,[14] and IDO is a molecular signature of tolerogenic DCs.[15] Through localized tryptophan deficiency combined with the release of proapoptotic metabolites, DCs can thus exert an IDO-dependent homeostatic control over the proliferation and survival of peripheral T cells, and can promote antigen-specific tolerance.

Tα1 was found to induce IDO activity in DCs, and this affected Th priming and tolerization. IDO induction required TLR9 and type I IFN receptor signaling and resulted in IL-10 production.[16] The induction of IDO by Tα1 affected the relative ability of DCs to balance antigen-specific Th/T$_{reg}$ cell priming *in vitro* and *in vivo*. Tα1 increased priming for IFN-γ– and IL-10–producing CD4$^+$ T cells, that is, it induced Th1/T$_{reg}$ cell activation. IDO blockade prevented the activation of IL-10–producing CD4$^+$ cells, while sparing that of IFN-γ–producing cells, a finding confirming the causal link of IDO with priming for IL-10–producing T cells. Consistent with the finding that TLR9 stimulation can promote DC-mediated generation of CD4$^+$CD25$^+$ T$_{reg}$ cells,[17] CD4$^+$CD25$^+$ T cells induced by Tα1-treated DCs expressed FoxP3[9,16] that is crucially involved in the suppressive phenotype of T$_{reg}$ cells. Therefore, Tα1 induced CD4$^+$CD25$^+$ T$_{reg}$ cells through the activation of a TLR9-dependent pathway.

Boosting tolerance by Tα1

Within the new paradigm of resistance and tolerance mechanisms as two alternative but complementary host defense strategies,[18] boosting tissue tolerance is likely to be a useful strategy in infectious and noninfectious diseases. The mammalian intestinal and respiratory mucosal surfaces are continuously exposed to a complex and dynamic community of microorganisms. These microbes establish symbiotic relationships with their hosts, making important contributions to immune fitness and metabolism efficiency.[19,20] The immune system is pivotal in mediating the interactions between host and microbiota that shape the mucosal environment. At the mucosal surfaces, a system of tolerance and controlled inflammation to limit the response to dietary or pathogen-derived antigens is present. This regulation is achieved by a number of cell populations acting through a set of shared regulatory pathways.[21] Current knowledge holds that populations of suppressor T_{reg} cells constitute a pivotal mechanism of immunological tolerance. The potential role of malfunctioning T_{reg} cells in chronic inflammatory immune and autoimmune diseases is well documented. Learning how to successfully manipulate T_{reg} cell responses could result in more effective vaccines and immunomodulators. Indeed, dysfunction of the innate and adaptive immune systems associated with mucosa causes impairment of mucosal barrier function and development of localized or systemic inflammatory and autoimmune processes. When this complex system breaks down in a genetically predisposed individual, the resulting immune response my lead to localized and systemic inflammatory diseases.[22,23] By inducing IDO and promoting T_{reg} cells, Tα1 successfully ameliorated respiratory allergy and intestinal inflammation in different experimental models.[4,5] Thus, maintaining diplomatic relations between mammals and beneficial microbial communities could be added to the therapeutic and immunomodulatory properties of Tα1.

Tα1 promotes immunity and tolerance in hematopoietic transplantation

The reconstitution of immunity in the recipients remains the most challenging aspect of hematopoietic stem cell transplantation (HSCT). Successful T cell reconstitution needs to accommodate the induction of protective immunity to pathogens and tumors within the context of transplantation tolerance. This implicates the requirement of DCs capable of generating some form of dominant regulation that ultimately controls inflammation, pathogen immunity and tolerance. Tα1-primed DCs fulfill these multiple requirements, which include the induction of priming to pathogens and tolerance to alloantigens in experimental and human HSCT. In experimental HSCT, the infusion of Tα1-treated DCs resulted in pathogen clearance and prevention of graft-versus host disease (GVHD).[16] Thus, Tα1 acted as the ideal immunomodulator capable of antimicrobial priming and concomitant tolerance in HSCT. Accordingly, Tα1 has recently been shown to promote immune recipients of HLA-matched sibling T cell-depleted stem cell transplants in a phase I/II clinical trial presented by Ruggeri *et al.*[24] Remarkably, Tα1 also improved survival, as the cumulative incidence of nonrelapse mortality (NRM,

mainly infection-related) was 32% in controls versus 7% in Tα1-treated patients. Multivariable analyses that included diagnoses, disease status at transplant, conditioning regimen and donor lymphocyte infusions showed that Tα1 treatment was a significant independent factor predicting a lower incidence of NRM, which tended to provide better survival. Thus, Tα1 continues to live up to its promises.

Balancing immunity and tolerance: learning from Tα1

Through a multifaceted, pleiotropic immune activation, Tα1 appears to be a promising adjuvant candidate and is suitable for combination therapies aimed at the control of inflammation, immunity, and tolerance in a variety of clinical settings. In addition to living up to its promises, Tα1 continues to amaze us with its pleiotropy, multitargeting activity, and ability to utilize different molecular mechanisms depending on contexts. Heterocomplexes and collaborative interactions between endogenous and/or exogenous TLR agonists and Tα1 appears to lead to novel reciprocal trends of immune surveillance and effector mechanisms aiding safe disposal of pathogens and/or pathogenic insults whereby excessive inflammatory responses and tissue injury may effectively be prevented (Fig. 2). NF-κB, a family of seven structurally related transcription factors that play a central role in the stress response and inflammation[25,26] are likely exploited by Tα1 for its pleiotropic activity. Although the NF-κB subunits are ubiquitously expressed, their actions are regulated in a cell type- and stimulus-specific manner, allowing for a diverse spectrum of effects. Molecular dissection of its mechanisms for activation has shown that NF-κB can be induced by the "canonical" and "noncanonical" pathways, leading to distinct patterns in the individual subunits activated and downstream genetic responses produced. Much attention has been focused on the proinflammatory signaling of canonical NF-κB, but data also indicate that noncanonical NF-κB could have opposing roles, limiting canonical NF-κB activity, inducing IDO, and controlling inflammation (Fig. 2).[27] Thus, the intersection between canonical and noncanonical NF-κB signaling pathways and IRFs may be crucial in promoting an optimally protective response balanced between inflammation and tolerance by Tα1.

Acknowledgments

We thank Dr. Cristina Massi Benedetti for digital art and editing. The original studies conducted in the author's laboratory were supported by the Italian Projects AIDS 2010 by the Istituto Superiore di Sanità (contract number 40H40 to L.R.) and by the Specific Targeted Research Project ALLFUN (FP7-HEALTH-2009 contract number 260338 to L.R.).

Conflicts of interest

The authors declare no conflicts of interest.

References

1. Goldstein, A.L. 2009. From lab to bedside: emerging clinical applications of thymosin alpha 1. *Expert. Opin. Biol. Ther.* **9:** 593–608.
2. Garaci, E. *et al.* 2000. Thymosin alpha 1 in the treatment of cancer: from basic research to clinical application. *Int. J. Immunopharmacol.* **22:** 1067–1076.
3. Billich, A. 2002. Thymosin alpha1. SciClone Pharmaceuticals. *Curr. Opin. Investig. Drugs* **3:** 698–707.
4. Pierluigi, B. *et al.* 2010. Thymosin alpha1: the regulator of regulators? *Ann. N.Y. Acad. Sci.* **1194:** 1–5.
5. Romani, L. *et al.* 2007. Thymosin alpha1: an endogenous regulator of inflammation, immunity, and tolerance. *Ann. N.Y. Acad. Sci.* **1112:** 326–338.
6. Steinman, R.M., D. Hawiger & M.C. Nussenzweig. 2003. Tolerogenic dendritic cells. *Annu. Rev. Immunol.* **21:** 685–711.
7. Kawai, T. & Akira, S. 2011. Toll-like receptors and their crosstalk with other innate receptors in infection and immunity. *Immunity* **34:** 637–650.
8. Kabelitz, D. 2007. Expression and function of Toll-like receptors in T lymphocytes. *Curr. Opin. Immunol.* **19:** 39–45.
9. Romani, L. *et al.* 2004. Thymosin alpha 1 activates dendritic cells for antifungal Th1 resistance through toll-like receptor signaling. *Blood* **103:** 4232–4239.
10. Zhang, P. *et al.* 2005. Activation of IKK by thymosin alpha1 requires the TRAF6 signalling pathway. *EMBO Rep.* **6:** 531–537.
11. Bozza, S. *et al.* 2007. Thymosin alpha1 activates TLR9/MyD88/IRF7-dependent murine cytomegalovirus sensing for induction of anti-viral responses *in vivo.* *Int. Immunol.* **19:** 1261–1270.
12. Grohmann, U., F. Fallarino & P. Puccetti. 2003. Tolerance, DCs and tryptophan: much ado about IDO. *Trends Immunol.* **24:** 242–248.
13. Mellor, A.L. & D.H. Munn. 2004. IDO expression by dendritic cells: tolerance and tryptophan catabolism. *Nat. Rev. Immunol.* **4:** 762–774.
14. Fallarino, F. & P. Puccetti. 2005. Toll-like receptor 9-mediated induction of the immunosuppressive pathway of tryptophan catabolism. *Eur. J. Immunol.* **36:** 8–11.
15. Orabona, C. *et al.* 2006. Toward the identification of a tolerogenic signature in IDO-competent dendritic cells. *Blood* **107:** 2846–2854.
16. Romani, L. *et al.* 2006. Thymosin alpha1 activates dendritic cell tryptophan catabolism and establishes a regulatory environment for balance of inflammation and tolerance. *Blood* **108:** 2265–2274.
17. Moseman, E.A. *et al.* 2004. Human plasmacytoid dendritic cells activated by CpG oligodeoxynucleotides induce the generation of CD4+CD25+ regulatory T cells. *J. Immunol.* **173:** 4433–4442.
18. Medzhitov, R., D.S. Schneider & M.P. Soares. 2012. Disease tolerance as a defense strategy. *Science* **335:** 936–941.
19. Hooper, L.V. & A.J. Macpherson. 2010. Immune adaptations that maintain homeostasis with the intestinal microbiota. *Nat. Rev. Immunol.* **10:** 159–169.
20. Huang, Y.J. & S.V. Lynch. 2011. The emerging relationship between the airway microbiota and chronic respiratory disease: clinical implications. *Expert. Rev. Respir. Med.* **5:** 809–821.
21. Saraiva, M. & A. O'Garra. 2010. The regulation of IL-10 production by immune cells. *Nat. Rev. Immunol.* **10:** 170–181.
22. Clemente, J.C. *et al.* 2012. The impact of the gut microbiota on human health: an integrative view. *Cell* **148:** 1258–1270.
23. Cho, I. & M.J. Blaser. 2012. The human microbiome: at the interface of health and disease. *Nat. Rev. Genet.* **13:** 260–270.
24. Perruccio, K. *et al.* 2010. Thymosin alpha1 to harness immunity to pathogens after haploidentical hematopoietic transplantation. *Ann. N.Y. Acad. Sci.* **1194:** 153–161.
25. Hayden, M.S. & S. Ghosh. 2004. Signaling to NF-kappaB. *Genes Dev.* **18:** 2195–2224.
26. Bonizzi, G. & M. Karin. 2004. The two NF-kappaB activation pathways and their role in innate and adaptive immunity. *Trends Immunol.* **25:** 280–288.
27. Grohmann, U. *et al.* 2007. Reverse signaling through GITR ligand enables dexamethasone to activate IDO in allergy. *Nat. Med.* **13:** 579–586.

Ann. N.Y. Acad. Sci. ISSN 0077-8923

ANNALS OF THE NEW YORK ACADEMY OF SCIENCES

Issue: *Thymosins in Health and Disease*

Single-domain β-thymosins: the family history

John Edwards

School of Life Sciences, University of Glasgow, Glasgow, United Kingdom

Address for correspondence: John Edwards, Honorary Research Fellow, School of Life Sciences, University of Glasgow, 154, Mugdock Road, Milngavie, Glasgow, G62 8NE. john@tannoch.org

Evolution probably invented the β-thymosin domain in a single-celled close relative of multicellular animals. Expansion from single genes to the small family of monomeric β-thymosins of present-day vertebrates may have started with a very ancient duplication, before the rounds of whole-genome duplications. In land vertebrates and fish, this family consists of the descendants of five genes of their jawed vertebrate common ancestor. Identifying this common ancestry depends on the genes possessing conserved sets of flanking sequences, as the relationships are not recognizable from amino-acid sequences. One of these genes has given rise both to a group of fish β-thymosins and to a hitherto unrecognized group of β-thymosins of birds and reptiles. The resulting classification may prove useful in relation to the β-thymosins of model organisms, such as the zebrafish, and for identifying important noncoding sequence elements, exemplified here by a conserved sequence in the 3′untranslated region of transcripts from the β4 subfamily.

Keywords: β-thymosin; synteny; teleost; tetrapod; whole genome duplication

Introduction

As thymosin β4 was discovered in a quest for hormones produced by the thymus,[1] it was a surprise to find that it serves as an important intracellular sequestering protein for G-actin.[2] It has since been found in a wide range of different animal tissues,[3] often very highly expressed. Great interest now focuses on its ability to moderate inflammation[4–6] and to promote regeneration of a variety of tissues,[7] offering promise of therapeutic applications to major human diseases such as heart attack and stroke, as recounted elsewhere in this volume.

The duality between cytoskeletal and regenerative roles, probably intra- and extracellular, respectively, has led to the use of the term *moonlighting* for the latter.[8] The diversity of activities of thymosin β4 may depend on its ability as an intrinsically unstructured protein to bind promiscuously to several different unrelated proteins.[9]

To help understand these activities, and particularly their enigmatic diversity, it may be useful to trace the evolution of the gene, and its relationship with other members of the β-thymosin family. The ever-accelerating flood of nucleotide sequences from multiple genome projects pro-

vides a rich source of data for such an undertaking. β-thymosins are not very well served by the automated annotation associated with large-scale genomics, and so there remains a role for hand-curated informatics.

Here, I summarize a classification of vertebrate β-thymosins based on ancestry of their genes, together with a model of the family history. I then outline two examples of directions in which such context may contribute to future understanding of the activities of thymosin β4 and its relatives.

Exploring the family relationships

β-thymosins as proteins of multicellular animals

β-thymosin domains, whether in ~40 amino-acid monomers, or in the tandem repeat proteins derived from them,[10] belong to a subset of protein domains found exclusively in the group now named *holozoans*—multicellular animals together with a few of their closest single-celled relatives, choanoflagellates and filastereans.[11–13] Amino-acid sequences retrieved from nucleotide sequence databases of β-thymosins from three species of single-celled organisms are shown in Figure 1. β-thymosins appear

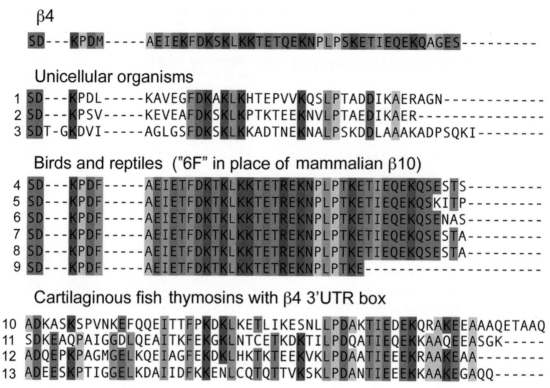

β4

SD---KPDM-----AEIEKFDKSKLKKTETQEKNPLPSKETIEQEKQAGES---------

Unicellular organisms

1 SD---KPDL-----KAVEGFDKAKLKHTEPVVKQSLPTADDIKAERAGN-----------
2 SD---KPSV-----KEVEAFDKSKLKPTKTEEKNVLPTAEDIKAER---------------
3 SDT-GKDVI-----AGLGSFDKSKLKKADTNEKNALPSKDDLAAAKADPSQKI--------

Birds and reptiles ("6F" in place of mammalian β10)

4 SD---KPDF-----AEIETFDKTKLKKTETREKNPLPTKETIEQEKQSESTS---------
5 SD---KPDF-----AEIETFDKTKLKKTETREKNPLPTKETIEQEKQSKITP---------
6 SD---KPDF-----AEIETFDKTKLKKTETREKNPLPTKETIEQEKQSENAS---------
7 SD---KPDF-----AEIETFDKTKLKKTETREKNPLPTKETIEQEKQSESTA---------
8 SD---KPDF-----AEIETFDKTKLKKTETREKNPLPTKETIEQEKQSESTA---------
9 SD---KPDF-----AEIETFDKTKLKKTETREKNPLPTKE-------------------

Cartilaginous fish thymosins with β4 3'UTR box

10 ADKASKSPVNKEFQQEITTFPKDKLKETLIKESNLLPDAKTIEDEKQRAKEEAAAQETAAQ
11 SDKEAQPAIGGDLQEAITKFEKGKLNTCETKDKTILPDQATIEQEKKAAQEEASGK-----
12 ADQEPKPAGMGELKQEIAGFEKDKLHKTKTEEKVKLPDAATIEEEKRAAKEAA--------
13 ADEESKPTIGGELKDAIIDFKKENLCQTQTTVKSKLPDANTIEEEKKAAKEGAQQ------

Figure 1. Novel β-thymosins referred to in the text. For evidence and accessions, see Supporting Information, Table S1. Residues are Taylor-coloured where % identity in group >70%: 1, *Monosiga brevicollis*; 2, *Salpingoeca* sp.; 3, *Capsaspora owczarzaki*; 4, Chicken; 5, Lizard; 6, Turkey; 7, Zebrafinch; 8, Budgerigar; 9, Python; 10, Cloudy catshark; 11, Little skate; 12, Dogfish shark; 13, Pacific electric ray.

to be absent from more distantly related eukaryotes, such as fungi and amoebae.[14]

This straightforward phylogeny emerges only if a distinction is drawn between β-thymosin and Wiskott-Aldrich Homology 2 (WH2) domains. Despite the close similarity in conformation of WH2 domains and the N-terminal half of β-thymosins when each is bound to actin—the "β-thymosin/WH2 fold,"[15] the two are distinguishable by conserved differences in amino-acid sequences,[14,16] and they are distinct genetically (unpublished). This is not to discount the possibility that WH2 domains may have been ancestral to β-thymosins. Contrary to some accounts,[10,17] the tandem repeat proteins (actobindins) of *Acanthamoebae* are constructed from WH2, rather than from β-thymosin domains.[14,16]

The multiple β-thymosins of land vertebrates
Although invertebrates express a single monomeric β-thymosin, vertebrates express several.[3,18–20] β-thymosins have a precise nomenclature appropriate

for small peptides, in which a unique number has been assigned to each if it differs from others by more than a few amino-acid residues.[20] Those of land vertebrates can be grouped by sequence similarity, forming three main subfamilies containing, respectively, the human proteins, thymosins β4, β10, and β15.[19,21]

Avian/reptilian "6F:" the fourth β-thymosin of land vertebrates
At least 13 mammals for which assembled genomes are available have genes for members of all three subfamilies, and transcript libraries where available confirm that all three are expressed. (β4 and β10 are read from single genes, whereas many placental mammals transcribe both members of a pair of β15 genes.)[19,22] However, although birds and reptiles also have members of the β4 and β15 subfamilies, in these animals β10 (or the very similar β9) is replaced with a different β-thymosin.[21] All six currently available sequences of this reptilian protein (Fig. 1) have residue 6 = phenylalanine (F), and

so here it is referred to as "6F." Expression of 6F in zebrafinch and lizard is supported by numerous expressed sequence tags (ESTs). Although the 6F protein is very similar to β4 in amino-acid sequence, the gene has a more recent common ancestor with β10 (see below).

Genomics indicates that these four subfamilies: β4, β10/β9, β15, and 6F are likely to complete the catalogue of land vertebrate β-thymosins.

The β-thymosins of teleost fish

No study of a vertebrate protein family can ignore teleost fish, which, with an estimated 22,000 species, reputedly outnumber all other vertebrates.[23]

The availability of five assembled teleost genomes provides a secure basis for a classification comparable with that outlined above for land vertebrates: it emerges that again there are four subfamilies. Conserved synteny (see below) largely confirms the subfamily relationships obtained from amino-acid similarity.[21] Members of two of these subfamilies are very similar in amino-acid sequence prompting the numbering 1A, 1B, 2, and 3.[21] Amino-acid similarity then allows assignment to these subfamilies of β-thymosins from many other teleost species for which assembled genomes are lacking.[21]

Common ancestry between tetrapods and teleosts

There is now consensus that the gene families of vertebrates were expanded during evolution by rounds of whole genome duplication (WGD): two en route to land vertebrates (tetrapods), and a third specific to teleost fish.[24–26] Many of the duplicated copies were not retained.[27] Because the first two rounds predated divergence of teleosts and tetrapods, it should be possible to trace the teleost and tetrapod β-thymosin subfamilies to shared ancestral genes of their gnathostome (jawed vertebrate) common ancestor. This would amount to determining which, in the terminology introduced by Fitch,[28] are the teleost *orthologs* of β4, β10, β15, and 6F.

Determination of orthology generally makes much use of similarity of amino-acid sequences. For β-thymosins, similarity allows them to be assigned to subfamilies *within* the two major evolutionary branches: tetrapods and teleosts. However, it conspicuously fails at the greater evolutionary distance needed to identify the orthology *between* these two branches. For β-thymosins, the amino-acid variations that differentiate the separate

tetrapod and teleost subfamilies are not correlated between the two branches (Fig. 2).

Fortunately, teleost/tetrapod orthology can be ascertained from conserved synteny. This works because β-thymosin genes, in common with many other vertebrate genes, preserve a fairly stable chromosomal hinterland (the identity and to some extent order, of flanking genes) through long evolutionary distances.

The outcome is summarized in Figure 2. Conserved synteny confirms the subfamilies arrived at from amino-acid sequence similarity within tetrapods and teleosts.[21] It also shows that 6F of birds and reptiles is a distinct subfamily in that it shares chromosomal context with none of the other three of land vertebrates. Importantly, it reveals clear common ancestry, despite their markedly discordant amino-acid sequences, between β4 and teleost group 1A, 6F and teleost group 2, and β15 and teleost group 3. These groups are outlined in black in Figure 2. Further information (see below) is needed to clarify how tetrapod β10-thymosins fit into this four-subfamily scheme.

The twin teleost orthologs of β4

Teleost groups 1A and 1B are sibling copies resulting from the teleost-specific third WGD. This is indicated by the gray box in Figure 2 and emerges from their respective chromosomal locations, which are on pairs known to result from the teleost-specific third WGD (e.g., medaka chromosomes 2 and 21, tetraodon 2 and 3).[25] This both explains and is confirmed by the absence of a 1B thymosin from zebrafish, a species which is known to have dispensed with the entire WGD3 sibling chromosome corresponding to its chromosome 9 which harbors thymosin 1A.[29] The β4/1A/1B relationship of the other four teleosts is an example of "doubly-conserved synteny."[30] However, synteny between 1B and β4 is weaker than between 1A and β4 (unpublished), indicating that the chromosomal segment harboring the B copy has experienced significantly more rearrangement following WGD3 than the stable context of 1A.

The family history deduced from a reconstruction of ancestral vertebrate chromosomes

Based chiefly on the genomes of human, chicken, and the teleost medaka, Nakatani et al.[31] defined a set of stable segments of vertebrate chromosomes,

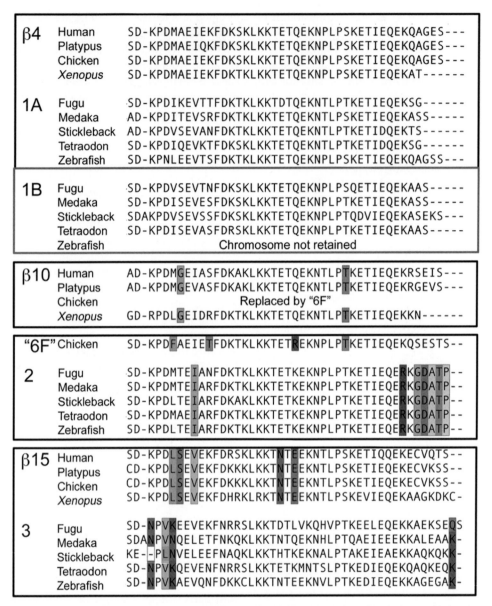

Figure 2. A gene-based classification of vertebrate β-thymosins. Species included are those for which assembled genomes were available: all five available teleosts and a representative range of land vertebrates. Residues are Taylor-colored to highlight sequence differences from 1A/1B for teleosts, and from β4 for land vertebrates. The stickleback group 3 amino-acid sequence is aberrant, and there is no evidence that the gene is transcribed. Sequences outlined with black boxes share flanking sequences, and so are believed to be descended from the same gene in the last common ancestor of teleosts and land vertebrates.[21] The gray outline reflects the weaker synteny of 1B teleosts with 1A and land vertebrates, interpreted as the 1A/1B pair having originated in the teleost-specific third round of whole genome duplication (WGD3).

referred to as conserved vertebrate linkage (CVL) blocks. These could be located in the chromosomes of the three species and used to extrapolate theoretical linkage groups for common ancestors at several stages in vertebrate evolution. Their paper included, in Supporting information, an atlas showing the mapping of over 100 CVL blocks in the three species. Using this data, the CVL blocks corresponding with the loci of β-thymosin genes can easily be identified. This yielded a model (Fig. 3) of how the

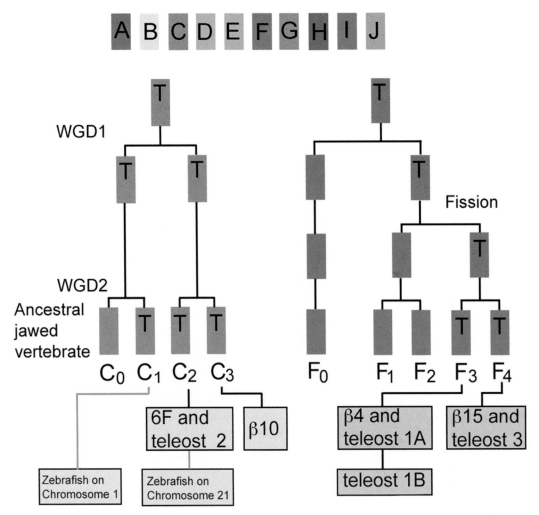

Figure 3. Model of the evolutionary history of vertebrate β-thymosins. In part, redrawn from Nakatani *et al.*,[31] on whose reconstruction of vertebrate ancestral chromosomes (shown as colored rectangles A–J) the model is based. Descendent chromosomes are labeled with "T" where extrapolation from present-day chromosomes indicates presence of a β-thymosin gene. C_0–C_3 and F_0–F_4 are reconstructed chromosomes of the jawed vertebrate common ancestor of tetrapods and teleosts. Possible origin of the additional thymosin genes of the zebrafish is indicated in smaller font.

β-thymosin family may have expanded in relation to WGDs.[21] It indicates that the β4/ β15 pair diverged at WGD2 and, importantly, that 6F and β10 subfamilies are similarly related. Such pairs are known as ohnologs, in honor of Ohno who first proposed a role for whole-genome duplications in vertebrate evolution.[32]

Unexpectedly, however, the respective CVL blocks containing these four genes extrapolate to *two different protochromosomes* in the vertebrate ancestor, indicating that a very ancient pre-WGD and presumably local segmental duplication of a β-thymosin gene began the family expansion.

"Ohnologs gone missing"

The flanking sequences of teleost group 2 thymosins, besides mapping around 6F (for example, on chicken chromosome 13) also identify segments of conserved synteny on mammalian chromosomes but which in these animals are devoid of β-thymosin genes, (e.g., on human chromosome 5).[33]

Thus, in the light of the Nakatani reconstruction, mammals, and probably amphibians[21] have retained one ohnolog, β10/ β9, and birds and reptiles a different one, 6F, and each lineage has dispensed with the alternate copy.

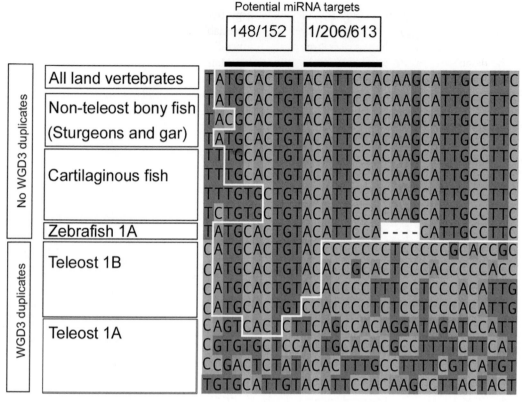

Figure 4. The thymosin β4 3′ UTR box. Conserved sequences found ~20 residues downstream of the stop codon in respective β-thymosin transcripts were hand-aligned, except for the group 1A teleosts, in which the sequence is not conserved, so a segment of an alignment of entire 3′ UTRs derived using Muscle[45] was substituted. The white line highlights segments of partial conservation. The two examples of possible target motifs for binding micro-RNAs were identified by TargetScan.[46] For accessions and further details, see Supporting information.

Some uses for the family history

The β-thymosins of zebrafish

In view particularly of its ability to regenerate functional heart tissue after amputation of up to 20% of the ventricle,[34–36] the zebrafish is an important model organism in relation to understanding possible developmental and regenerative activities of β-thymosins.

Zebrafish express five β-thymosins, in comparison with the three of the other teleosts for which assembled genomes are available. In common with the others, zebrafish share a member of each of β4/1A, 6F/2, and β15/3 subfamilies, but differ in lacking the WGD3 second (1B) β4 ortholog—in this respect, a better model for mammalians. However, zebrafish express two additional β-thymosins, which synteny indicates may be, respectively: a WGD3

duplicate of the 6F/2 gene and a probable WGD2 copy of 6F/2 not retained by other vertebrates[21] (Fig. 3). Although three of the five have featured in experimental studies, surprisingly this apparently does not yet include the actual β4 ortholog on chromosome 9. For further details of the five, see Table S2.

The conserved β4 3′ untranslated region (UTR) box

One respect in which the interspecies gene relationships may aid understanding of β4 follows from the power of conservation of noncoding sequences to highlight important functions—the approach known as phylogenetic footprinting,[37,36] (chiefly applied to promoters). This involves the aligning of sequences from orthologous genes in different species.

An early discovery was that a sequence in the 3′ UTR of thymosin β4 is highly conserved in mammals.[38] Subsequently, the vastly enlarged nucleotide databases provide confirmation that this 32-residue sequence is conserved with total identity in β4 subfamily thymosins of at least 32 species of land vertebrates, from primates to amphibians. Because it is highly specific for the β4 orthologs, I refer to it here as the thymosin β4 3′ UTR box. Database searches for this sequence also recover a set of β-thymosin monomer sequences from non-teleost bony fish (sturgeons and gar). Remarkably, a sequence with 26–30 of its residues identical is present in β-thymosin transcripts from cartilaginous fishes (Fig. 4). The latter have strikingly different amino-acid sequences (Fig. 1), and should synteny (via the little skate genome[39]) in the future confirm the presumption that these too are β4 orthologs, this will provide a further vivid illustration of the conspicuous lack of correlation of β-thymosin–coding sequences with orthology.

The motif is however *not* conserved in the teleost β4 orthologs other than partially in zebrafish 1A (Fig. 4). This may be a consequence of the 1A/1B WGD3 duplication, because all the vertebrates (tetrapods and nonteleost fish) in which the box is conserved have not experienced a WGD3, and the zebrafish is unique (among the five teleosts for which there are assembled genomes) in lacking a 1B ortholog. This may be an instance of subfunctionalization, which is postulated as driving retention of duplicated genes.[40,41]

The function of the 3′ UTR box is unknown, but it is likely to be relevant that it contains several candidate target sequences for interaction with micro-RNAs and thus potentially for developmentally important posttranslational regulation.[42,43] Interestingly, one of these targets, TGCACTG *is* conserved in the four group 1B teleost thymosins. Thus comparisons among fish group 1 β-thymosins may assist analysis of 3′UTR function in human β4, in terms of independent smaller motifs.

Conclusion

The consistent patterns of difference of amino-acid sequence, which define β-thymosin subfamilies within the major vertebrate branches, tetrapods and teleosts are believed to fine-tune the functions of these proteins.[18,44] However, whereas these patterns are not correlated between the major branches, respective chromosomal contexts are shared to an appreciable extent. Such contexts, together with noncoding regions of transcripts are likely to be important in the developmental programming of expression. The relationships described here should provide useful input to future experimental work as well as stimulus to further β-thymosin–focused bioinformatics.

Acknowledgments

I thank numerous genome-sequencing and assembly groups for sequence assemblies. These include the Bovine Genome Project at Baylor College of Medicine; the Broad Institute, Genome Sequencing Center of Washington University in St. Louis; Genoscope, Japanese National Institute of Genetics and University of Tokyo; the National Center for Biotechnology Information; the U.S. DOE Joint Genome Institute; and the Wellcome Trust Sanger Institute.

I thank Robin Stevenson, Tony Lawrence, and John Young for provoking my interest in β-thymosins, and Allan Goldstein for sustaining it.

Supporting Information

Additional supporting information may be found in the online version of this article.

Table S1. Evidence, Pfam Expect values, and GenInfo identifiers for novel β-thymosins referred to in the text. Amino-acid sequences are in Figure 1.

Table S2. The five β-thymosins expressed by zebrafish *Danio rerio*

Table S3. Species and Geninfo identifiers for β4 3′UTR box alignment in Figure 4

Please note: Wiley-Blackwell is not responsible for the content or functionality of any supporting materials supplied by the authors. Any queries (other than missing material) should be directed to the corresponding author for the article.

Conflicts of interest

The author declares no conflicts of interest.

References

1. Low, T.L.K., S.K. Hu & A.L. Goldstein. 1981. Complete amino acid sequence of bovine thymosin β4: a thymic hormone that induces terminal deoxynucleotidyl transferase activity in thymocyte populations. *Proc. Natl. Acad. Sci. USA* **78**: 1166–1170.

2. Safer, D., M. Elzinga & V.T. Nachmias. 1991. Thymosin beta 4 and Fx, an actin-sequestering peptide, are indistinguishable. *J Biol. Chem.* **266:** 4029–4032.

3. Huff, T., C.S. Müller, A.M. Otto, *et al.* 2001. Beta-thymosins, small acidic peptides with multiple functions. *Int. J. Biochem. Cell Biol.* **33:** 205–220.

4. Young, J.D., A.J. Lawrence, A.G. MacLean, *et al.* 1999. Thymosin beta 4 sulfoxide is an anti-inflammatory agent generated by monocytes in the presence of glucocorticoids. *Nat. Med.* **5:** 1424–1427.

5. Sosne, G., E.A. Szliter, R. Barrett, *et al.* 2002. Thymosin beta 4 promotes corneal wound healing and decreases inflammation *in vivo* following alkali injury. *Exp. Eye Res.* **74:** 293–9.

6. Badamchian, M. 2003. Thymosin beta 4 reduces lethality and down-regulates inflammatory mediators in endotoxin-induced septic shock. *Int. Immunopharmacol.* **3:** 1225–1233.

7. Philp, D. & H.K. Kleinman. 2010. Animal studies with thymosin β4, a multifunctional tissue repair and regeneration peptide. *Ann. N. Y. Acad. Sci.* **1194:** 81–86.

8. Goldstein, A.L., E.Hannappel & H.K. Kleinman. 2005. Thymosin beta4: actin-sequestering protein moonlights to repair injured tissues. *Trends Mol. Med.* **11:** 421–429.

9. Tompa, P., C. Szász & L. Buday. 2005. Structural disorder throws new light on moonlighting. *Trends Biochem. Sci.* **30:** 484–489.

10. Van Troys, M., S. Dhaese, J. Vandekerckhove, *et al.* 2007. Multirepeat β-thymosins. In *Actin-Monomer-Binding Proteins.* Lappalainen, P., Ed. Springer, New York.

11. King, N., M.J. Westbrook, S.L. Young, *et al.* 2007. The genome of the choanoflagellate *Monosiga brevicollis* and the origin of metazoans. *Nature* **451:** 783–788.

12. Lang, B.F., C. O'Kelly, T. Nerad, *et al.* 2002. The closest unicellular relatives of animals. *Curr. Biol.* **12** 1773–1778.

13. Shalchian-Tabrizi, K., M.A. Minge, M. Espelund, *et al.* 2008. Multigene phylogeny of *Choanozoa* and the origin of animals. *PLoS ONE* **3:** e2098.

14. Punta, M., P.C. Coggill, R.Y. Eberhardt, *et al.* 2012. The Pfam protein families database. *Nucleic Acids Res.* Database Issue **40:** D290–D301.

15. Dominguez, R. 2007. The beta-thymosin/WH2 fold: multifunctionality and structure. *Ann. N. Y. Acad. Sci.* **1112:** 86–94.

16. Edwards, J. 2004. Are β-thymosins WH2 domains? *FEBS Lett.* **573:** 231–232.

17. Hertzog, M., E.G. Yarmola, D. Didry, *et al.* 2002. Control of actin dynamics by proteins made of beta-thymosin repeats: the actobindin family. *J. Biol. Chem.* **277:** 14786–14792.

18. Eadie, J.S., S.W. Kim, P.G. Allen, *et al.* 2000. C-terminal variations in beta-thymosin family members specify functional differences in actin-binding properties. *J. Cell Biochem.* **77:** 277–287.

19. Dhaese, S., K. Vandepoele, D. Waterschoot, *et al.* 2009. The mouse thymosin beta15 gene family displays unique complexity and encodes a functional thymosin repeat. *J. Mol. Biol.* **387:** 809–825.

20. Hannappel, E. 2010. Thymosin beta4 and its posttranslational modifications. *Ann. N. Y. Acad. Sci.* **1194:** 27–35.

21. Edwards, J. 2010. Vertebrate β-thymosins: conserved synteny reveals the relationship between those of bony fish and of land vertebrates. *FEBS Lett.* **584:** 1047–1053.

22. Banyard, J., L.M. Hutchinson & B.R. Zetter. 2007. Thymosin β-NB is the human isoform of rat thymosin β15. *Ann. N.Y. Acad. Sci.* **1112:** 286–296.

23. Taylor, J.S., I. Braasch, T. Frickey, *et al.* 2003. Genome duplication, a trait shared by 22000 species of ray-finned fish. *Genome Res.* **13:** 382–390.

24. Dehal, P. & J.L. Boore. 2005. Two rounds of whole genome duplication in the ancestral vertebrate. *PLoS Biol.* **10:** e314.

25. Meyer, A. & Y. Van de Peer. 2005. From 2R to 3R: evidence for a fish-specific genome duplication. *Bioessays* **27:** 937–945.

26. Kasahara, M. 2007. The 2R hypothesis: an update. *Curr. Opin. Immunol.* **19:** 547–552.

27. Huminiecki, L. & C.H. Heldin. 2010. 2R and remodeling of vertebrate signal transduction engine. *BMC Biol.* **8:** 146.

28. Fitch, W.M. 1970. Distinguishing homologous from analogous proteins. *Syst. Zool.* **19:** 99–113.

29. Kasahara, M., K. Naruse, S. Sasaki, *et al.* 2007. The medaka draft genome and insights into vertebrate genome evolution. *Nature* **447:** 714–719.

30. Kellis, M., B.W. Birren & E.S. Lander. 2004. Proof and evolutionary analysis of ancient genome duplication in the yeast Saccharomyces cerevisiae. *Nature* **428:** 617–624.

31. Nakatani, Y., H. Takeda, Y. Kohara, *et al.* 2007. Reconstruction of the vertebrate ancestral genome reveals dynamic genome reorganization in early vertebrates. *Genome Res.* **17:** 1254–1265.

32. Ohno, S. 1970. *Evolution by Gene Duplication.* Springer-Verlag, New York.

33. Postlethwait, J.H. 2007. The zebrafish genome in context: ohnologs gone missing. *J. Exp. Zool. B Mol. Dev. Evol.* **308:** 563–577.

34. Poss, K.D., L.G. Wilson & M.T. Keating. 2002. Heart regeneration in zebrafish. *Science* **298:** 2188–2190.

35. Zhang, Z. & M. Gerstein. 2003. Of mice and men: phylogenetic footprinting aids the discovery of regulatory elements. *J. Biol.* **2:** 11–11.4.

36. Jopling, C., E. Sleep, M. Raya, *et al.* 2010. Zebrafish heart regeneration occurs by cardiomyocyte dedifferentiation and proliferation. *Nature* **464:** 606–609.

37. Xie, X., J. Lu, E.J. Kulbokas, *et al.* Systematic discovery of regulatory motifs in human promoters and 3′ UTRs by comparison of several mammals. *Nature* **434:** 338–345.

38. Duret, L. & P. Bucher. 1997. Searching for regulatory elements in human noncoding sequences. *Curr. Opin. Struct. Biol.* **7:** 399–406.

39. Wang, Q., C.N. Arighi, B.L. King, *et al.* 2012. Community annotation and bioinformatics workforce development in concert—Little Skate Genome Annotation Workshops and Jamborees. *Database.* doi: 10.1093/database/bar064.

40. Lynch, M. & A. Force. 2000. The probability of duplicate gene preservation by subfunctionalization. *Genetics* **154:** 459–473.

41. Braasch, I. & W. Salzburger. 2009. *In ovo omnia:* diversification by duplication in fish and other vertebrates. *J. Biol.* **8:** 25.

42. Bartel, D.P. 2004. MicroRNAs: genomics, biogenesis, mechanism, and function. *Cell* **116:** 281–97.

43. Fabian, M.R., N. Sonenberg & W. Filipowicz. 2010. Regulation of mRNA translation and stability by microRNAs. *Ann. Rev. Biochem.* **79:** 351–379.

44. Jean, C., K. Rieger, L. Blanchoin, *et al.* 1994. Interaction of G-actin with thymosin beta 4 and its variants thymosin beta 9 and thymosin beta met9. *J. Muscle Res. Cell Motil.* **15:** 278–286

45. Edgar, R.C. 2004. MUSCLE: multiple sequence alignment with high accuracy and high throughput. *Nucleic Acids Res.* **32:** 1792–1797.

46. Friedman, R.C., K.K. Farh, C.B. Burge, *et al.* 2008. Most mammalian mRNAs are conserved targets of microRNAs. *Genome Res.* **19:** 92–105.

Ann. N.Y. Acad. Sci. ISSN 0077-8923

Controlled delivery of thymosin β4 for tissue engineering and cardiac regenerative medicine

Loraine L.Y. Chiu,[1] Lewis A. Reis,[2] and Milica Radisic[1,2,3]

[1]Chemical Engineering and Applied Chemistry, University of Toronto, Toronto, Canada. [2]Institute of Biomaterials and Biomedical Engineering, University of Toronto, Toronto, Canada. [3]Heart and Stroke/Richard Lewar Center of Excellence, University of Toronto, Toronto, Canada

Address for correspondence: Dr. Milica Radisic, Associate Professor, Institute of Biomaterials and Biomedical Engineering, Department of Chemical Engineering and Applied Chemistry, Heart and Stroke/Richard Lewar Center of Excellence, University of Toronto, 164 College St., Rm. 407, Toronto, Ontario, M5S 3G9 Canada. milica@chem-eng.utoronto.ca

Thymosin β4 (Tβ4) is a peptide with multiple biological functions. Here, we focus on the role of Tβ4 in vascularization, and review our studies of the controlled delivery of Tβ4 through its incorporation in biomaterials. Tβ4 promotes vascularization through VEGF induction and AcSDKP-induced migration and differentiation of endothelial cells. We developed a collagen–chitosan hydrogel for the controlled release of Tβ4 over 28 days. *In vitro*, the Tβ4-encapsulated hydrogel increased migration of endothelial cells and tube formation from epicardial explants that were cultivated on top of the hydrogel, compared to Tβ4-free hydrogel and soluble Tβ4 in the culture medium. *In vivo*, subcutaneously injected Tβ4-containing collagen–chitosan hydrogel in rats led to enhanced vascularization compared to Tβ4-free hydrogel and collagen hydrogel with Tβ4. Furthermore, the injection of the Tβ4-encapsulated hydrogel in the infarct region improved angiogenesis, reduced tissue loss, and retained left ventricular wall thickness after myocardial infarction in rats.

Keywords: thymosin β4; peptide; heart; angiogenesis; hydrogel

Introduction

Cardiac tissue engineering and regenerative medicine is often complicated by the need for vascularization to provide oxygen and nutrients within highly metabolic cardiac tissues.[1,2] To generate tissue-engineered cardiac patches, cardiomyocytes were seeded on scaffolds and cultivated in bioreactors. Because, oxygen concentration is greatly reduced near the center of cardiac constructs due to the high oxygen consumption rates by cardiomyocytes, leaving a 100 μm thick outer layer of viable cells with a nearly cell-free interior,[3–5] different bioreactors have been used to improve the transport of oxygen for cultivation of cardiac constructs *in vitro*, including perfusion bioreactors[3–4] and rotating bioreactors.[6] Recent approaches for cardiac tissue engineering recognize the need for *in vitro* vasculogenesis and cardiac protection by the fabrication and use of bioinstructive biomaterials incorporating proangiogenic bioactive molecules, such as vascular endothelial growth factor (VEGF),[7–10] basic fibroblast growth factor,[11] and small peptides like RGD.[12]

Thymosin β4 (Tβ4) is a small 43-amino acid polypeptide that has various biological roles, including actin binding and sequestration, induction of angiogenesis, and cardioprotection. Tβ4 was identified as an angiogenic factor by an *in vitro* screening of early genes (< 4 h) induced during cultivation of human umbilical vein endothelial cells (HUVECs) on Matrigel^{TM}.[13] Tβ4 promotes vessel formation and collateral growth during development.[14] Moreover, exogenous Tβ4 enhanced tube formation of HUVECs *in vitro* and vascular sprouting in the coronary artery ring angiogenesis assay.[15] Its role in vascular development motivated research by several groups to investigate its biological potential in the adult organisms, as discussed here.

doi: 10.1111/j.1749-6632.2012.06718.x

Role of Tβ4 in cardioprotection and neovasculogenesis

Tβ4 was reported to be important for cardioprotection and cardiac repair. When injected intraperitoneally or intracardially, Tβ4 improves cardiac repair after coronary artery ligation, with significantly higher fractional shortening and ejection fraction at four weeks after myocardial infarction (MI).[16] In addition, the scar size was reduced with Tβ4 treatment due to early protection of the myocardium through Tβ4-mediated cardiomyocyte survival.

A possible mechanism of Tβ4-induced cardioprotection is the stimulation of coronary neovascularization. Mouse embryos express Tβ4 in the cardiovascular system, including the left ventricle, outer curvature of the right ventricle and cardiac outflow tract.[16] In addition, Tβ4 has been found to play a role in restoring the pluripotency of the otherwise quiescent adult epicardium.[14] Tβ4 treatment induced outgrowth from epicardial explants from hearts of 8- to 12-week-old mice, while untreated explants showed minimal outgrowth. Tβ4 directly promotes epicardial cell migration through its actin binding, filament assembly, and lamellipodia formation. The outgrowth cell population was positive for epicardin—an epicardial-specific transcription factor—and could be differentiated into fibroblasts, smooth muscle cells, and endothelial cells. This is a significant finding, since the supply of endothelial and smooth muscle vascular precursors required for vascular regeneration was previously attributed to the bone marrow and peripheral circulation. The potential to use Tβ4 to release quiescent epicardial cells as a source of vascular progenitors suggests the ability to deliver endothelial and smooth muscle cells to the sites of interest following cardiac injury for vascular repair.

A recent study by Smart *et al.*[17] demonstrated that pretreatment of the adult heart with Tβ4 was important for activating Wilm's tumor 1-expressing (WT-1$^+$) epicardial cells that contributes to both vasculogenesis and cardiomyogenesis after MI. However, if Tβ4 was applied at the time of MI or after MI there was no effect of Tβ4 on epicardial cell plasticity.[18] Pretreatment with Tβ4 prior to MI is not feasible in clinical settings, and systemic application must be carefully evaluated, since Tβ4 can cause growth of cancerous lesions in the gut.[19]

Thus, formulation of a controlled delivery system that can sustain the effect of Tβ4 even at low doses is required.

In a study by Qian *et al.*,[20] local delivery of Gata4, Mef2c, and Tbx5 (GMT) into the boundary between the infarct and border zones converted resident cardiac fibroblasts into cardiomyocyte-like cells *in vivo* after cardiac injury. The co-delivery of Tβ4 with GMT further improved cardiac function *in vivo*, as it activated the proliferation of cells and in turn increased the delivery of GMT to more cells. In larger animals, controlled delivery, by placing the Tβ4-containing gel onto the epicardium, may be beneficial because this will require less peptide, and the activity of the peptide can be localized to the epicardium, as required.

Mechanisms of action

While Tβ4 has many functions, its mechanism of action has not been completely defined. It was previously shown that Tβ4 binds to cells and becomes internalized,[5] for example, radioiodinated Tβ4 was found to bind to the cell surface of HUVECs. In addition, incubation of exogenous Tβ4 increased intracellular staining of Tβ4. It is thought that the increase of intracellular Tβ4 could be due either to uptake of Tβ4 by the cells or to a response to exogenous Tβ4 to increase transcription and translation of Tβ4 mRNA. The internalization of Tβ4 was shown to be necessary for antiapoptotic activity in human corneal epithelial cells.[21] It was not until recently that an extracellular signaling pathway was identified for Tβ4, in which Tβ4 increases cell surface ATP levels by binding to the β subunit of ATP synthase.[22] The same study showed that the ATP-responsive P2X4 receptor was required for Tβ4 to induce migration of HUVECs.

A study performed to determine the portion of Tβ4 responsible for its angiogenic activity showed that the actin binding domain of Tβ4, LKTET, is essential for endothelial cell adhesion and angiogenesis.[23] The addition of exogenous actin blocked adhesion of HUVECs and vessel sprouting in the aortic ring assay mediated by Tβ4, showing that actin is the cell surface ligand for Tβ4. Various synthetic peptides and naturally occurring proteolytic fragments of Tβ4 have been used in cell migration and vessel sprouting assays, and it was determined that the peptides containing the carboxyl-terminal portion or all of the actin binding domain promoted cell migration and vessel sprouting. A small 7-amino acid synthetic peptide containing amino acids 17–23 from Tβ4, LKKTETQ, promoted migration and sprouting, and blocked adhesion to Tβ4. Based on such data, Tβ4 is thought to initiate angiogenesis through the binding of its central actin binding domain to the surface actin of endothelial cells. However, a separate study showed that the addition of synthetic peptide with actin binding sequence LKK-TETQEK did not compete with Tβ4 to bind to the cell surface of HUVECs.[15] This suggests that the binding of Tβ4 does not occur through the actin binding domain.

It has also been shown that Tβ4 stimulates angiogenesis through induction of VEGF, a potent angiogenic growth factor. Overexpression of Tβ4 was found to upregulate VEGF in mouse melanoma cells.[24] In addition, the knockdown of Tβ4 downregulated VEGF in mouse hearts.[14] Tβ4 has also been shown to upregulate the levels of hypoxia-inducible factor (HIF)-1α in mouse melanoma B16-F10 cells. Further, Jo *et al.*[25] found that Tβ4 induced the transcriptional expression of VEGF through the stabilization of HIF-1α protein.

In addition, the proangiogenic tetrapeptide *N*-acetyl-seryl-aspartyl-lysyl-proline (AcSDKP) was found to be an bioactive fragment formed by cleavage of Tβ4 at its N terminus between Pro 4 and Asp 5 by prolyl oligopeptidase.[14,26] AcSDKP levels were significantly reduced after Tβ4 knockdown.[14] AcSDKP is proangiogenic, stimulating the proliferation and migration of endothelial cells and the formation of capillary-like structures on Matrigel by these cells.[27,28] AcSDKP induced angiogenesis *in vivo* in the rat Matrigel plug angiogenesis assay,[27] chick chorioallantoic membrane assay,[27] rat corneal micropocket assay,[28] and rat myocardial infarction model.[28] In the adult mouse, there was upregulation of both Tβ4 and AcSDKP after ischemia.[14] AcSDKP did not have an actin-binding function and thus could not promote outgrowth from epicardial explant as Tβ4 did. AcSDKP-treated cells differentiated into Flk-positive endothelial cells but not smooth muscle cells, suggesting that AcS-DKP only supports endothelial cell differentiation. Thus, the vasculogenic effect of Tβ4 may be attributed to this fragment. AcSDKP may also have immunomodulatory properties, as AcSDKP-induced vascularization after hind-limb ischemia by

induction of monocyte chemoattractant protein-1 (MCP-1).[29]

Besides improving cardiac repair through inducing vascularization, Tβ4 also acts through the activation of Akt. Akt is a serine/threonine kinase that plays a role in cell survival and angiogenesis.[30] It can be activated by phosphorylation at Ser 473 by kinases such as integrin-linked kinase (ILK).[31] Tβ4 was shown to form a functional complex with particularly interesting Cys-His protein (PINCH) and ILK,[16] in turn activating Akt for cell survival. There was an increase in the level of ILK protein and phosphorylated Akt-S473 in mouse hearts with Tβ4 treatment after coronary ligation, compared with PBS treatment.[16] Thus, activation of ILK, which in turn stimulates Akt activation, may be one mechanism by which Tβ4 enhances the survival of cardiomyocytes. However, other experiments showed that Tβ4-knockdown in mice led to high levels of phosphorylated Akt-S473.[14] This suggests that the activation of Akt may either be a secondary response to Tβ4 treatment or a compensatory mechanism for the lack of Tβ4.

In vitro *application and* in vivo *delivery of Tβ4*

The potential for Tβ4 as a therapeutic for treatment of cardiac diseases has been established, yet a full understanding of the mechanism and scope of its potential use is only now being explored. As the internalization of Tβ4 may be required for its effects, there are limitations of the modes in which it can be administered both *in vitro* and *in vivo*. Logically, *in vitro* studies have been focused on using soluble Tβ4 in cell media to study vasculogenesis, cell migration, and differentiation; and most *in vivo* studies have done the same by using Tβ4 in soluble form (reviewed in Ref. 32).

Excitement over the discovery of the angiogenic and cardioprotective effects of Tβ4 has led to animal studies and human clinical trials involving the systemic administration of Tβ4 as a treatment for acute MI.[14,16,33] Bock-Marquette *et al.*[16] treated acute MI through the injection of Tβ4 in mice; the peptide was delivered systemically through intraperitoneal injections of Tβ4 in PBS every three days or locally within the cardiac infarct through intracardial injection of Tβ4 in a collagen gel once only. Another delivery regime used was a combination of initial intracardial injection followed by intraperitoneal injections every three days for four weeks. Systemic, intracardial or a combination of both systemic and intracardial injections of Tβ4 in the mouse acute MI model showed that treatment with Tβ4 led to a greater than 60% improvement in fractional shortening and more than 100% improvement in ejection fraction compared to PBS injection controls.[16] Furthermore, cardiac morphology was significantly improved in Tβ4-treated animals compared to controls. However, there was no difference between systemic or local delivery of Tβ4 in this case.

A phase I human clinical trial involved determining the safety and tolerability of single and multiple ascending intravenous injections of soluble Tβ4 in healthy volunteers.[33] Successful completion of the study has paved the way for approval to move to phase II trials, giving soluble intravenous Tβ4 to acute MI patients (http://clinicaltrials. gov/ct2/show/NCT01311518). While animal studies showed no difference in improvement between systemic or local delivery, and human trials confirmed the safety of systemic administration with large dose Tβ4 injections, there is still interest in investigating local delivery of Tβ4 to prevent washout of the peptide and to improve efficacy of the peptide at the cardiac infarct.

The development of controlled delivery systems for Tβ4 is motivated by the need to prolong and localize the bioactivity of peptide. Injectable hydrogels capable of sustained, localized delivery of bioactive molecules were studied in many different applications.[34,35] Only a few groups are looking at the development of such hydrogels in direct conjunction with Tβ4 specifically. Collaboration between the Langer and Hubbell groups (from Massachusetts Institute of Technology in the United States and Ecole Polytechnique Federale de Lausanne in Switzerland, respectively) has led to the development of a polyethylene glycol (PEG)-based hydrogel modified with matrix metalloproteinase (MMPs) cleavable peptide domains.[36] As such, the hydrogel was prone to cell-mediated proteolytic degradation and remodeling. Tβ4 was physically entrapped in the hydrogel, with gelling of the hydrogel occurring in approximately 30 min at physiological conditions. Addition of MMP-2 and MMP-9 (at concentrations of 100 ng/mL and 1000 ng/mL) resulted in hydrogel degradation and release of close to 100% of entrapped Tβ4 within ~170 hours for MMP-2 and ~190 hours for MMP-9. HUVECs en-

capsulated within the developed system exhibited increased survival in hydrogels containing Tβ4, as well as enhanced MMP-2 and MMP-9 secretion in the presence of increasing concentrations of Tβ4 (up to 40 mg/mL). Furthermore, Tβ4 improved HUVEC attachment to the hydrogel and induced vascular-like networks to be formed within the matrix.[37]

In vivo, injection of the PEG hydrogel in a rat MI model showed 80% of encapsulated Tβ4 release by day 3 after injection and 95% release by six weeks.[37] The gel degraded to about 25% of the injected amount by day 28 and was undetectable at six weeks. Injection of the gel (MMP+PEG) significantly reduced infarct size compared to a PBS control showing that the gel alone can enhance cardiac function. However, addition of Tβ4 with human embryonic stem cell (hESC)-derived vascular cells further improved host cardiomyocyte functionality in the infarct zone. A modest improvement in cardiac function could be seen with the combined treatment after six weeks, with an increase in ejection fraction of ~12% with Tβ4 and hESC-derived cell treatment compared to PBS injection control. There was also more vasculature in the treatment group and transplanted vascular cells were found to form *de novo* vascular structures. They hypothesized that the gel substituted for the degrading endogenous matrix, providing temporary support and pro-survival factors, while hESC-derived vascular cells contributed to formation of capillary-like vessels, stabilization of host vessels, secretion of paracrine factors, or induction of paracrine factor secretion from native rat cells.

Collagen–chitosan hydrogel for controlled delivery of Tβ4

We developed a collagen–chitosan hydrogel capable of controlled delivery of Tβ4 (Fig. 1).[38] This hydrogel was composed of a 2:1 collagen to chitosan ratio. Tβ4 was mixed with the hydrogel solution and became encapsulated upon gelation. The hydrogel could deliver Tβ4 in a controlled manner over 28 days, with a burst release of 18–28% over the initial three days and approximately zero order kinetics from day 3 to 28 (Fig. 1A). In contrast, hydrogel composed of collagen alone released the full Tβ4 load (1491 ± 27 ng) after the first three days. The collagen–chitosan hydrogel had a porous structure and gelation time of 200–300 seconds. The

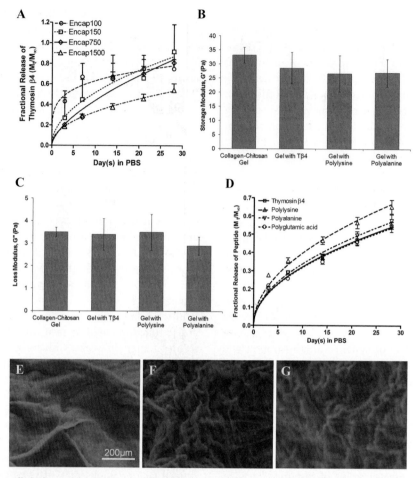

Figure 1. Controlled release of peptides from collagen–chitosan composite hydrogel.[38] The developed collagen–chitosan hydrogel was used for controlled release of different peptides, including the negatively charged Tβ4 (∼11 nm in length) and polyglutamic acid (∼7 nm in length), neutral polyalanine (∼14 nm in length), and positive-charged polylysine (∼5 nm in length). (A) Cumulative release profiles of Tβ4 with different initially encapsulated Tβ4 amounts, shown as a fraction of initial loading released over time. Groups include hydrogel with 100 ng (Encap100), 150 ng (Encap150), 750 ng (Encap750), and 1,500 ng (Encap1500) encapsulated Tβ4 in 50 μL. Release studies were performed in PBS at 37 °C ($n = 6$–8 samples/group). (B) Storage modulus of collagen–chitosan hydrogel with or without encapsulated peptides. Groups include hydrogel with no peptide (collagen–chitosan gel), hydrogel with encapsulated Tβ4 (Gel with Tβ4), hydrogel with encapsulated polylysine (gel with polylysine), and hydrogel with encapsulated polyalanine (gel with polyalanine). (C) Loss modulus of collagen–chitosan hydrogel with or without encapsulated peptides. (D) Cumulative release profiles of different peptides, shown as a fraction of initial loading released over time (1,500 ng peptide encapsulated). (E–G) Scanning electron microscopy images of (E) collagen-only gel, (F) collagen–chitosan composite hydrogel, and (G) composite hydrogel with 1,500 ng encapsulated Tβ4. See Ref. 38.

plateau loss and storage moduli were determined to be approximately 3 Pa and 30 Pa, respectively (Fig. 1B, C). Since loss modulus (representing viscous properties) was lower than storage modulus (representing elastic properties), elastic properties were dominating, and a stable gel was formed. The encapsulation of peptides did not change gelation time or mechanical properties. We found that the hydrogel was capable of release of neutral, posi-

tively charged, and negatively charged peptides over 28 days (Fig. 1D) based on diffusive mechanism, according to the fit of their release profiles to the Ritger–Peppas equation, where $n = 0.5$ indicates diffusive mechanism. This may be related to the porous structure of collagen–chitosan hydrogel, as shown by scanning electron microscopy, compared to a smoother structure in collagen-only hydrogel (Fig. 1E–G). The release of charged peptides

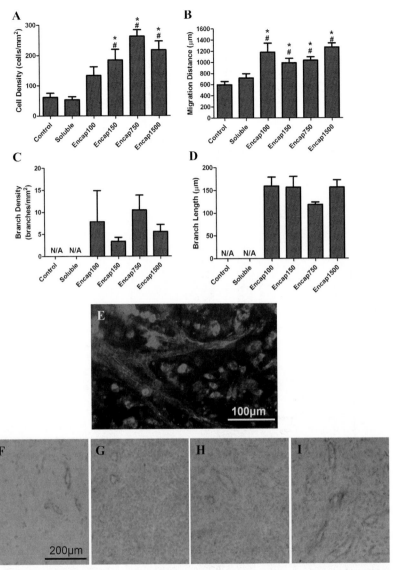

Figure 2. Controlled delivery of Tβ4 by incorporation into collagen–chitosan composite hydrogel led to increased cell outgrowths and tube formation from cultivated epicardial tissue explants, as well as angiogenesis when subcutaneously injected.[38] (A–E) YFP mouse cardiac explants were placed onto the hydrogel with different amounts of Tβ4, and fluorescence microscopy images were taken after seven days of *in vitro* cultivation ($n = 8$–12 samples/group). Migrated cell density, cell migration distance, branch density, and branch length were measured from the images using ImageJ. (A) Density of migrated cells from adult mouse cardiac explants. (B) Maximum distance of cell migration from the cardiac explants. (C) Density of branches sprouted from cardiac explants. (D) Average length of branches that sprouted from cardiac explants. N/A, no sprouting observed. Groups include hydrogels without Tβ4 (Control), cultivation with soluble Tβ4 in culture medium (Soluble), or hydrogel with 100 ng (Encap100), 150 ng (Encap150), 750 ng (Encap750), and 1,500 ng (Encap1500) encapsulated Tβ4. 50 µL hydrogels were used. Cardiac explants were cultivated for seven days. *Statistically significant difference compared to control. #Statistically significant difference compared to soluble ($P < 0.05$ one-way ANOVA with *post hoc* Tukey tests). (E) Fluorescence microscopy image showing CD31 immunostaining of Tβ4-induced sprouting from cardiac explants cultivated on hydrogel with 1500 ng encapsulated Tβ4 (red indicates CD31-positive staining of tubes formed by endothelial cells, blue indicates staining of the nuclei). This demonstrates that the capillary-like tubes were formed by the migrated endothelial cells, due to the angiogenic activity of Tβ4. Cells that migrated from the cardiac explants on hydrogels with 1,500 ng encapsulated Tβ4 included 31% endothelial cells, 33% fibroblasts, 9% smooth muscle cells, and 27% myofibroblasts. (F–I) Hydrogels, 100 µL, were subcutaneously injected in rats, and factor VIII staining after seven days showed increased angiogenesis with the injection of (I) collagen–chitosan hydrogel with 1,500 ng Tβ4 compared to (F) collagen hydrogel with 1,500 ng Tβ4, (G) bare collagen–chitosan hydrogel, and (H) hydrogel with a lower dose of 100 ng Tβ4. See Ref. 38.

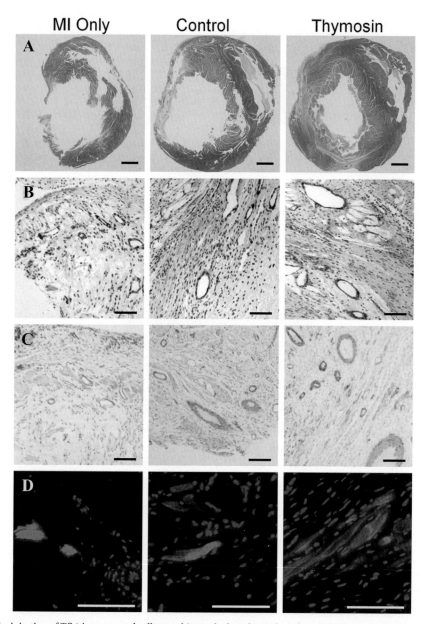

Figure 3. The injection of Tβ4-incorporated collagen–chitosan hydrogel into the infarct region promoted angiogenesis, reduced tissue loss and attenuated left ventricular wall thinning post-MI.[42] (A) Masson's trichrome staining of heart sections showing a thicker left ventricular wall with the injection of Tβ4 hydrogel (scale = 1 mm). (B) Factor VIII staining showing higher blood vessel density and larger vessel diameter within the infarct region with Tβ4 treatment (scale = 100 μm). (C) There was higher number of smooth muscle actin-positive blood vessels with larger vessel diameter within the infarct region with Tβ4 treatment (scale = 100 μm). (D) Troponin-T staining showing enhanced cardiomyocyte presence in the infarct region with Tβ4 treatment (scale = 100 μm, red indicates positive staining of cardiomyocytes, blue indicates staining of nuclei). Groups include no treatment (MI Only), treatment with 50 μL bare collagen–chitosan hydrogel (control), and treatment with 50 μL collagen–chitosan hydrogel with 1,500 ng Tβ4 (thymosin) after left anterior descending artery ligation in rats. The hydrogels were injected immediately after MI in all experiments ($n = 3$–5 animals/group). Data are from Ref. 38.

deviated from purely diffusive mechanism, likely due to the electrostatic interactions between the charged peptides and the hydrogel. Collagen is negatively charged and chitosan is positively charged at neutral pH, but collagen–chitosan blends at various collagen to chitosan ratios were previously found to be positively charged.[39]

Upon developing the system that successfully sustained the release of Tβ4, it was necessary to demonstrate that the peptide is still bioactive. The controlled release of Tβ4 enhanced cell outgrowths from adult mouse cardiac explants that were cultivated on top of the developed collagen–chitosan hydrogels (Fig. 2).[38] There was an increased number of cells that migrated out of the cultivated explants with the controlled delivery of Tβ4, compared to explants grown on plain hydrogel or cultured with soluble Tβ4 (Fig. 2A, $P < 0.0001$). Importantly, 4,500 ng Tβ4 was used in total for the soluble case, while the use of 150 ng to 1,500 ng encapsulated Tβ4 was sufficient to significantly increase cell migration from the tissue explants. The cell migration distance was also increased with the controlled release of Tβ4, compared to Tβ4-free and soluble Tβ4 controls (Fig. 2B, $P < 0.0001$). There were no cardiac troponin-T$^+$ cells found in the population of migrated cells. Importantly, the growth of mouse epicardial explants on Tβ4-encapsulated hydrogel led to capillary outgrowths, while this was not observed without Tβ4 or with soluble Tβ4 (Fig. 2C, D). The capillary outgrowths were composed of CD31-positive endothelial cells (Fig. 2E).

The developed hydrogel was then evaluated *in vivo* by a subcutaneous injection into rats. Degradation of the hydrogel was determined by injecting biotinylated collagen–chitosan hydrogel subcutaneously and then quantifying biotin content of the hydrogel samples after extraction from the rats. There was no difference in biotin content of the hydrogel samples from day 0 and day 7, thus indicating no degradation after seven days. Controlled release of Tβ4 enhanced vascularization after seven days *in vivo*, compared to injection of hydrogel alone or collagen-only hydrogel with incorporated Tβ4 (Fig. 2F–I). The vessel density was 6.7 ± 0.6 vessels/mm^2 in collagen–chitosan hydrogel with 1,500 ng Tβ4, compared to only 0.4 ± 0.1 vessels/mm^2 for collagen gel with 1,500 ng Tβ4, 4.4 ± 0.5 vessels/mm^2 for Tβ4-free collagen–chitosan hy-

drogel, and 4.9 ± 0.6 vessels/mm^2 for collagen–chitosan hydrogel with 100 ng Tβ4.

It was then necessary to evaluate the importance of controlled delivery of Tβ4 with the developed hydrogel in a MI model. The injection of biomaterials has been shown to prevent pathological remodeling.[40,41] Wall *et al.*[41] used a finite element model simulation of the damaged ovine left ventricular wall to show that injection of hydrogels alone in the border zone can prevent pathological remodeling due to mechanical effects such as reduction of elevated myofiber stresses. Two alginate materials are now in phase II clinical trials for treatment of heart failure (ClinicalTrials.gov Identifiers: NCT01226563, NCT01311791). Functionalizing these hydrogels with bioactive molecules could provide an additional effect. In our study, the Tβ4-encapsulated hydrogel was injected into the infarct region after performing left anterior descending artery (LAD) ligation on adult rats (Fig. 3).[42] After three weeks, there was reduced wall thinning and tissue loss with the injection of Tβ4-encapsulated hydrogel compared to hydrogel alone. The left ventricular wall thickness was 1.07 ± 0.08 mm for treatment with Tβ4 hydrogel compared to only 0.27 ± 0.03 mm for no treatment and 0.77 ± 0.11 mm for treatment with Tβ4-free hydrogel (Fig. 3A, $P < 0.001$ for thymosin versus MI only; $P < 0.05$ for thymosin versus control). The average blood vessel diameter was increased within the infarct region (Fig. 3B, C, $P = 0.0169$ for factor VIII, $P = 0.0077$ for smooth muscle actin). The average FVIII-positive blood vessel diameter was 17 ± 1 μm for treatment with Tβ4 hydrogel compared to only 12 ± 2 μm for no treatment and 15 ± 1 μm for treatment with Tβ4-free hydrogel. The average smooth muscle actin-positive blood vessel diameter was 19 ± 1 μm for treatment with Tβ4 hydrogel compared to only 12 ± 1 μm for no treatment and 17 ± 1 μm for treatment with Tβ4-free hydrogel. In addition, there was enhanced cardiomyocyte presence in the infarct area after Tβ4 treatment (Fig. 3D, $P < 0.0001$). The percentage of troponin T$^+$ cardiomyocytes within the infarct area was $15 \pm 3\%$ for treatment with Tβ4 hydrogel compared to only $4 \pm 2\%$ for no treatment and $8 \pm 3\%$ for treatment with Tβ4-free hydrogel. The combination of blood vessel maturation and cardiomyocyte presence in the infarct area may have reduced the left ventricular wall thinning in our study. In turn, maintaining a thick left ventricular wall is important for preventing an increase in wall stress that can lead to heart failure.

The advantages of the developed collagen–chitosan hydrogel here compared to the aforementioned PEG hydrogels with MMP cleavable domains are as follows: (a) MMPs can cleave Tβ4 therefore decreasing its activity while collagen–chitosan hydrogel does not require action of MMPs for controlled release of the encapsulated peptides; and (b) PEG hydrogels with MMP cleavable domains were demonstrated to retain Tβ4 for a period of 72–200 h, while collagen–chitosan hydrogels can sustain the release for a period of four weeks, a time frame relevant to the process of the pathological remodeling of the heart upon myocardial infarction.

Conclusions

In this paper we discussed the role of Tβ4 in adult neovasculogenesis and cardioprotection and its mechanisms of action. We also presented delivery modes of Tβ4 for cardiac repair, with a focus on our studies involving the encapsulation of Tβ4 in collagen–chitosan composite hydrogels. While clinical trials are ongoing using systemic delivery of soluble Tβ4, it is clear that controlled delivery can improve the efficacy of the peptide by providing a localized and sustained source of the peptide. The various functions of Tβ4 motivate further research in developing controlled release systems for the peptide.

Conflicts of interest

The authors declare no conflicts of interest.

References

1. Morritt, A.N., S.K. Bortolotto, R.J. Dilley, *et al.* 2007. Cardiac tissue engineering in an *in vivo* vascularized chamber. *Circulation* 115: 353–360.
2. Caspi, O., A. Lesman, Y. Basevitch, *et al.* 2007. Tissue engineering of vascularized cardiac muscle from human embryonic stem cells. *Circ. Res.* 100: 263–272.
3. Carrier, R.L., M. Rupnick, R. Langer, *et al.* 2002. Perfusion improves tissue architecture of engineered cardiac muscle. *Tissue Eng.* 8: 175–188.
4. Radisic, M., A. Marsano, R. Maidhof, *et al.* 2008. Cardiac tissue engineering using perfusion bioreactor systems. *Nat. Protoc.* 3: 719–738.
5. Radisic, M., J. Malda, E. Epping, *et al.* 2006. Oxygen gradients correlate with cell density and cell viability in engineered cardiac tissue. *Biotechnol. Bioeng.* 93: 332–343.
6. Bursac, N., M. Papadaki, J.A. White, *et al.* 2003. Cultivation in rotating bioreactors promotes maintenance of cardiac myocyte electrophysiology and molecular properties. *Tissue Eng.* 9: 1243–1253.
7. Odedra, D., L.L. Chiu, M. Shoichet & M. Radisic. 2011. Endothelial cells guided by immobilized gradients of vascular endothelial growth factor on porous collagen scaffolds. *Acta Biomater.* 7: 3027–3035.
8. Chiu, L.L. & M. Radisic. 2010. Scaffolds with covalently immobilized VEGF and Angiopoietin-1 for vascularization of engineered tissues. *Biomaterials* 31: 226–241.
9. Miyagi, Y., L.L. Chiu, M. Cimini, *et al.* 2011. Biodegradable collagen patch with covalently immobilized VEGF for myocardial repair. *Biomaterials* 32: 1280–1290.
10. Wu, J., F. Zeng, X.P. Huang, *et al.* 2011. Infarct stabilization and cardiac repair with a VEGF-conjugated, injectable hydrogel. *Biomaterials* 32: 579–586.
11. Garbern, J.C., E. Minami, P.S. Stayton & C.E. Murry. 2011. Delivery of basic fibroblast growth factor with a pH-responsive, injectable hydrogel to improve angiogenesis in infarcted myocardium. *Biomaterials* 32: 2407–2416.
12. Yu, J., Y. Gu, K.T. Du, *et al.* 2009. The effect of injected RGD modified alginate on angiogenesis and left ventricular function in a chronic rat infarct model. *Biomaterials* 30: 751–756.
13. Grant, D.S., J.L. Kinsella, M.C. Kibbey, *et al.* 1995. Matrigel induces thymosin beta 4 gene in differentiating endothelial cells. *J. Cell. Sci.* 108(Pt 12): 3685–3694.
14. Smart, N., C.A. Risebro, A.A. Melville, *et al.* 2007. Thymosin beta4 induces adult epicardial progenitor mobilization and neovascularization. *Nature* 445: 177–182.
15. Grant, D.S., W. Rose, C. Yaen, *et al.* 1999. Thymosin beta4 enhances endothelial cell differentiation and angiogenesis. *Angiogenesis* 3: 125–135.
16. Bock-Marquette, I., A. Saxena, M.D. White, *et al.* 2004. Thymosin beta4 activates integrin-linked kinase and promotes cardiac cell migration, survival and cardiac repair. *Nature* 432: 466–472.
17. Smart, N., S. Bollini, K.N. Dube, *et al.* 2011. *De novo* cardiomyocytes from within the activated adult heart after injury. *Nature* 474: 640–644.
18. Zhou, B., L.B. Honor, Q. Ma, *et al.* 2012. Thymosin beta 4 treatment after myocardial infarction does not reprogram epicardial cells into cardiomyocytes. *J. Mol. Cell Cardiol.* 52: 43–47.
19. Tang, M.C., L.C. Chan, Y.C. Yeh, *et al.* 2011. Thymosin beta 4 induces colon cancer cell migration and clinical metastasis via enhancing ILK/IQGAP1/Rac1 signal transduction pathway. *Cancer Lett.* 308: 162–171.
20. Qian, L., Y. Huang, C.I. Spencer, *et al.* 2012. In vivo reprogramming of murine cardiac fibroblasts into induced cardiomyocytes. *Nature* 485: 593–598.
21. Ho, J.H., C.H. Chuang, C.Y. Ho, *et al.* 2007. Internalization is essential for the antiapoptotic effects of exogenous thymosin beta-4 on human corneal epithelial cells. *Invest. Ophthalmol. Vis. Sci.* 48: 27–33.
22. Freeman, K.W., B.R. Bowman & B.R. Zetter. 2011. Regenerative protein thymosin beta-4 is a novel regulator of purinergic signaling. *FASEB J.* 25: 907–915.

23. Philp, D., T. Huff, Y.S. Gho, *et al.* 2003. The actin binding site on thymosin beta4 promotes angiogenesis. *FASEB J.* **17:** 2103–2105.

24. Cha, H.J., M.J. Jeong & H.K. Kleinman. 2003. Role of thymosin beta4 in tumor metastasis and angiogenesis. *J. Natl. Cancer Inst.* **95:** 1674–1680.

25. Jo, J.O., S.R. Kim, M.K. Bae, *et al.* 2010. Thymosin beta4 induces the expression of vascular endothelial growth factor (VEGF) in a hypoxia-inducible factor (HIF)-1alpha-dependent manner. *Biochim. Biophys. Acta.* **1803:** 1244–1251.

26. Cavasin, M.A., N.E. Rhaleb, X.P. Yang & O.A. Carretero. 2004. Prolyl oligopeptidase is involved in release of the antifibrotic peptide Ac-SDKP. *Hypertension* **43:** 1140–1145.

27. Liu, J.M., F. Lawrence, M. Kovacevic, *et al.* 2003. The tetrapeptide AcSDKP, an inhibitor of primitive hematopoietic cell proliferation, induces angiogenesis *in vitro* and *in vivo. Blood* **101:** 3014–3020.

28. Wang, D., O.A. Carretero, X.Y. Yang, *et al.* 2004. N-acetyl-seryl-aspartyl-lysyl-proline stimulates angiogenesis *in vitro* and *in vivo. Am. J. Physiol. Heart Circ. Physiol.* **287:** H2099–H2105.

29. Waeckel, L., J. Bignon, J.M. Liu, *et al.* 2006. Tetrapeptide AcSDKP induces postischemic neovascularization through monocyte chemoattractant protein-1 signaling. *Arterioscler. Thromb. Vasc. Biol.* **26:** 773–779.

30. Brazil, D.P., J. Park & B.A. Hemmings. 2002. PKB binding proteins. Getting in on the Akt. *Cell* **111:** 293–303.

31. Rossdeutsch, A., N. Smart & P.R. Riley. 2008. Thymosin beta4 and Ac-SDKP: tools to mend a broken heart. *J. Mol. Med.* **86:** 29–35.

32. Crockford, D., N. Turjman, C. Allan & J. Angel. 2010. Thymosin beta4: structure, function, and biological properties supporting current and future clinical applications. *Ann. N.Y. Acad. Sci.* **1194:** 179–189.

33. Crockford, D. 2007. Development of thymosin beta4 for treatment of patients with ischemic heart disease. *Ann. N.Y. Acad. Sci.* **1112:** 385–395.

34. Nelson, D.M., Z. Ma, K.L. Fujimoto, *et al.* 2011. Intramyocardial biomaterial injection therapy in the treatment of heart failure: Materials, outcomes and challenges. *Acta Biomater.* **7:** 1–15.

35. Kretlow, J.D., L. Klouda & A.G. Mikos. 2007. Injectable matrices and scaffolds for drug delivery in tissue engineering. *Adv. Drug Deliv. Rev.* **59:** 263–273.

36. Kraehenbuehl, T.P., L.S. Ferreira, P. Zammaretti, *et al.* 2009. Cell-responsive hydrogel for encapsulation of vascular cells. *Biomaterials* **30:** 4318–4324.

37. Kraehenbuehl, T.P., L.S. Ferreira, A.M. Hayward, *et al.* 2011. Human embryonic stem cell-derived microvascular grafts for cardiac tissue preservation after myocardial infarction. *Biomaterials* **32:** 1102–1109.

38. Chiu, L.L. & M. Radisic. 2011. Controlled release of thymosin beta4 using collagen–chitosan composite hydrogels promotes epicardial cell migration and angiogenesis. *J. Control Release* **155:** 376–385.

39. Wang, X., L. Sang, D. Luo & X. Li. 2011. From collagen–chitosan blends to three-dimensional scaffolds: the influences of chitosan on collagen nanofibrillar structure and mechanical property. *Colloids. Surf. B. Biointerfaces* **82:** 233–240.

40. Leor, J., S. Tuvia, V. Guetta, *et al.* 2009. Intracoronary injection of in situ forming alginate hydrogel reverses left ventricular remodeling after myocardial infarction in Swine. *J. Am. Coll. Cardiol.* **54:** 1014–1023.

41. Wall, S.T., J.C. Walker, K.E. Healy, *et al.* 2006. Theoretical impact of the injection of material into the myocardium: a finite element model simulation. *Circulation* **114:** 2627–2635.

42. Chiu, L.L., L.A. Reis, A. Momen, & M. Radisic. 2012. Controlled release of thymosin beta4 from injected collagen-chitosan hydrogels promotes angiogenesis and prevents tissue loss after myocardial infarction. *Regen. Med.* **7:** 523–533.

Ann. N.Y. Acad. Sci. ISSN 0077-8923

ANNALS OF THE NEW YORK ACADEMY OF SCIENCES

Issue: *Thymosins in Health and Disease*

Thymosin α1 and cancer: action on immune effector and tumor target cells

Enrico Garaci,[1] Francesca Pica,[1] Annalucia Serafino,[2] Emanuela Balestrieri,[1] Claudia Matteucci,[1] Gabriella Moroni,[1] Roberta Sorrentino,[1] Manuela Zonfrillo,[2] Pasquale Pierimarchi,[2] and Paola Sinibaldi-Vallebona[1]

[1]Department of Experimental Medicine and Surgery, University of Rome "Tor Vergata," Rome, Italy. [2]Institute of Translational Pharmacology, National Research Council of Italy, Rome, Italy

Address for correspondence: Francesca Pica, M.D., PhD., Department of Experimental Medicine and Surgery, University of Rome "Tor Vergata", Via Montpellier, 1-00133, Rome, Italy. pica@uniroma2.it

Since it was first identified, thymosin alpha 1 (Tα1) has been characterized to have pleiotropic effects on several pathological conditions, in particular as a modulator of immune response and inflammation. Several properties exerted by Tα1 may be attributable to a direct action on lymphoid cells. Tα1 has been shown to exert an immune modulatory activity on both T cell and natural killer cell maturation and to have an effect on functions of mature lymphocytes, including stimulating cytokine production and cytotoxic T lymphocyte–mediated cytotoxic responses. In previous studies we have shown that Tα1 increases the expression of major histocompatibility complex class I surface molecules in murine and human tumor cell lines and in primary cultures of human macrophages. In the present paper, we describe preliminary data indicating that Tα1 is also capable of increasing the expression of tumor antigens in both experimental and human tumor cell lines. This effect, which is exerted at the level of the target tumor cells, represents an additional factor increasing the antitumor activity of Tα1.

Keywords: thymosin α1; cancer; immune effector cells; tumor antigens

Introduction

Thymosin α1 (Tα1) is a naturally occurring thymic peptide first described and characterized by Goldstein *et al.* in 1972.[1] In the form of a synthetic 28-amino acid peptide, Tα1 is in clinical trials worldwide for the treatment of various immune deficiencies, infectious diseases, and cancer. In addition, novel applications and other recently completed trials point to much broader clinical applications of Tα1 in the treatment of life-threatening and chronic diseases.[2]

Tα1 has proved to be capable of stimulating innate as well as adaptive immune responses in different experimental and human conditions of immune suppression. The first observations of the immune restorative effects of Tα1 in cancer therapy were by Chretien *et al.*[3] and Schulof *et al.*[4] in patients with lung cancer. These studies, showing an immune-restorative effect induced by Tα1 in patients undergoing radiotherapy, was an early proof-of-concept study for the efficacy of immunotherapeutic strategies. Since then, numerous experimental and clinical studies have tested the antitumor activity of Tα1, either used alone or in combination with cytokines and chemotherapeutic agents. The recognition that Tα1 is able to upregulate major histocompatibility complex (MHC) class-I antigen expression in normal and transformed cells,[5] and the herein reported evidence that it is also capable of influencing both differentiation and antigen expression in different tumor cells, point to an additional new important action of Tα1, which adds to its previously known ability of affecting lymphoid cells. All of these findings strongly encourage the further exploitation of Tα1 clinical use in cancer therapy.

Action on immune effector cells

Tα1 is known to be able of activating or boosting the immune system by several mechanisms, including stimulation of T cell differentiation and/or maturation[6,7], activation of natural killer (NK)[8] and

doi: 10.1111/j.1749-6632.2012.06697.x

Ann. N.Y. Acad. Sci. 1269 (2012) 26–33 © 2012 New York Academy of Sciences.

dendritic cells (DCs),[9] and induction of proinflammatory cytokine release.[10,11]

Multiple studies have demonstrated an effect of Tα1 on the production of a number of cytokines by peripheral blood lymphocytes (PBL), among which were migration inhibition factor (MIF),[12] colony stimulating factor (CSF),[13,14] granulocyte-macrophage CSF (GM-CSF), B cell growth factor,[15] interferon (IFN),[16–18] and interleukin-2 (IL-2).[19,20] For this last cytokine, the effect was even greater, as Leichtling *et al.* demonstrated that Tα1 was also capable of modulating the expression of high-affinity IL-2 receptors in mitogen-stimulated normal human lymphocytes.[20–22] Since then, Tα1 was thought to act not as a first signal in the induction of cytokine production but rather as a super-inducing accessory signal. Because of this characteristic, Tα1 appeared to have great potential when used with other cytokines in combination therapies.

Several years ago we showed that treatment with Tα1 for four days, followed by a single injection of α,β-IFN 24 hours before testing, strongly restored NK activity in cyclophosphamide (CY)-suppressed mice. Moreover, Tα1 was able to accelerate the recovery rate of NK activity in bone marrow–reconstituted murine chimeras. The data, taken together, support the concept that the synergistic effect between Tα1 and α,β-IFN could be the result of action on differentiation of the NK lineage at different levels.[8] NK cells are known to play a major role in the antitumor immune response. Thus, our approach to the treatment of tumors used a combined protocol consisting of the administration of Tα1 and IFN or IL-2 after CY. The rationale of the this triple combination was based on the assumption that chemotherapy, although reducing the tumor mass, induces a marked suppression of the immune response that could be restored by means of the combined immunotherapy.

This schedule of treatment resulted in a powerful antitumor effect in several preclinical models in mice (Lewis lung carcinoma, Friend erythroleukemia and B16 melanoma)[23–26] and in rats (DHD/K12 colorectal cancer [CRC] liver metastasis)[27] and cured a high percentage of treated tumor-bearing animals. Histological analysis of tumors explanted from animals, which had been treated with the triple combination therapy, showed a strong cellular infiltration by CD4, CD8, and

NK cells. The injection of monoclonal antibodies against these cells abrogated the curative effects of the protocol, confirming that the immune response was involved in eradicating the tumor.[23–26]

Subsequently, pivotal clinical trials in advanced nonsmall cell lung cancer (NSCLC)[28,29] and melanoma[30] using Tα1 and low dose IFN-α, following *cis*-platinum or dacarbazine (DTIC) respectively, indicated the feasibility of this approach in the treatment of human cancer. Finally, the results of a large randomized study, which was aimed at evaluating the efficacy and safety of combining Tα1 with DTIC and IFN-α in patients with metastatic melanoma, have confirmed that Tα1 may have beneficial effects on patient's survival, supporting further clinical evaluation of this agent.[31] Given the limited treatment options for this condition and the safety profile of Tα1, treatment with Tα1 could be considered assuming that the decision about its use was based on expert clinical judgement.[32]

As shown by other research groups, the relation between Tα1 and cancer is complex and based also on the direct effect of Tα1 on various immune effector cell types, among which are macrophages and DCs.[9,33–35] As an example, Shrivastava *et al.* have shown that Tα1 activates monocytes, bone marrow-derived macrophages (BMDMs) and tumor-associated macrophages to a tumoricidal state.[33] This effect occurs via the enhancement of their production of macrophage-activating cytokines and other molecules (i.e., IL-1, tumor necrosis factor (TNF), reactive oxygen intermediates (ROI), and nitric oxide (NO)), which act in autocrine manner, with consequent increased pinocytosis, phagocytosis, antigen presentation, and tumor cytotoxicity by these important effector cells.[33,34] Peng *et al.* have shown that Tα1 induces the production of IL-6, IL-10, and IL-12 in murine BMDMs through IκB kinase (IKK) and mitogen-activated protein kinases (MAPK) pathways.[35] Romani *et al.* have shown that Tα1 upregulates the expression of Toll-like receptor (TLR) 2, 5, 8, and 9 and protects mice from challenge by invasive aspergillosis in a myeloid differentiate factor 88 (MyD88)-dependent way.[9] Recent work by Romani *et al.* also shows that Tα1 fulfills the requirement of a promising adjuvant candidate for strategies aimed at the control of inflammation, immunity, and tolerance in a variety of clinical settings. Its regulatory function on DCs appears fundamental as DCs have

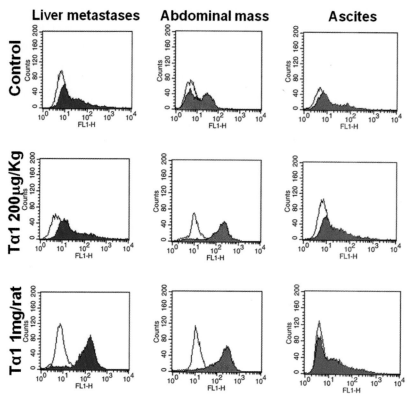

Figure 1. *In vivo* effect of Tα1 on CSH-275 antigen expression in CRC metastatic lesions induced in immunocompetent BDIX rats. DHD-K12 cells isolated from dissected liver and abdominal metastases and from ascitic fluid, obtained from treated and untreated animals, were subjected to indirect immunofluorescent staining using a rabbit polyclonal anti-CSH-275 revealed with a FITC-conjugated goat anti-rabbit IgG. The *ex vivo* samples were analyzed by flow cytometry. FITC-labeled rabbit preimmune serum was used for background measurements. The extent of antigen expression was calculated as a percentage of positive cells, reported in Table 1. These data represent work in progress.

a crucial role in determining immune outcomes by acquiring antigens, collating environmental cues, and becoming cells that are either potent stimulators or suppressors of T cell responses.[36,37]

The pleiotropic function of Tα1 is also well documented by our studies on the transcriptional response of lymphoid cells to this molecule, which have been performed by using gene-array technology.[38] By this experimental approach we demonstrated that Tα1 regulates the transcriptional response of a high number of genes involved in the immune response (cytotoxic activities [NK/CTL], IFN cascade, innate immune response, cytokines and chemokines, MHC, growth factors, and transcriptional factors).[38]

Despite the relevant body of evidence available to date, it appears that there is still a need for studies aimed at better understanding the molecular targets and pathways used by Tα1. Results of such studies

may lead to a better exploitation of its function in cancer as well as in other diseases.

Action on target tumor cells

Beginning in the 1980s, the improvement of knowledge of cancer biology and the development of biotechnology led to the identification and characterization of several new tumor-associated antigens (TAAs).[39,40] TAAs in the management of cancer patients, both as tumor markers for diagnosis or as targets for immunotherapy, still represent an area of intense investigation. In fact tumor antigen expression plays a major role in the success of immunotherapeutic protocols, and the new strategies for fighting tumors should be aimed both at increasing immune effector cell activities, through the modulation of cytokine expression, and at enhancing the expression of tumor cell antigens, thus rendering tumor cells more "visible" to immune effector cells and less prone to escape immune response. Antitumor immunity is primarily mediated by cytotoxic T lymphocytes (CTL), which recognize tumor antigens presented by MHC-I molecules.[39] The failure of immunotherapeutic protocols and of immune-based diagnostic methods in most cancers is mainly due to two concomitant factors occurring on tumor cells: the low intracellular expression of MHC class-I and the low expression of tumor antigens, which represent specific molecular targets. These events ultimately result in cancer cell escape from the host antitumor immunity. An adequate expression of tumor antigens is also needed for immune-diagnosis in early stages of the disease and for treatment of cancer cells with different immune-chemotherapeutic approaches.[39]

As indicated earlier, Tα1, when used alone or in combination with other agents, has been found to exert beneficial effects in cancer patients. Tα1 has also been reported to upregulate the expression of MHC class-I in not only lymphoid cells but murine and human tumor cell lines and in primary cultures of human macrophages, through a regulation of gene expression.[38] Tα1 was also found to be capable of inducing carcinoembryonic antigen (CEA) expression in WiDr CRC cells.[41] These findings point to a novel activity of Tα1, not only in relation to its effects on T lymphocytes or cytokine modulation but also at the target cell level, that is on tumor cells.

To expand our knowledge about the effects of Tα1 on TAA expression, in the last years we have performed a series of experiments using *in vitro* and *in vivo* experimental models of CRC and melanoma. The main results of these researches are reported herein.

More than ten years ago a new TAA, the non-apeptide epitope CSH-275 (RTNKEASIC), was identified and we demonstrated its expression in tissue specimens from human CRC, while normal colonic mucosa was found to be negative.[42] CSH-275 expression was also correlated with the sequential stages of human CRC transformation.[42] Subsequently, the CSH-275 epitope was found to be naturally expressed in the DHD-K12 rat colon adenocarcinoma cell line.[42,43] These cells, originally obtained from 1,2-dimethylhydrazine-induced colon adenocarcinoma in syngeneic immunocompetent BDIX rats, display many of the features of human epithelial cells constituting the CRC tissue, including not only the expression of the human TAA CSH-275[42,43] but also the constitutive activation of cancer-related pathways such as the Wnt/β-catenin pathway, as recently demonstrated.[44] The features sustain the usefulness of this cell line as a suitable *in vitro* model for developing therapeutic strategies for CRC.

When syngeneic BDIX rats were immunized with the CSH-275 peptide, CTL activity was successfully raised and DHD-K12 tumor cell growth *in vitro* was significantly reduced (Pierimarchi *et al.*, unpublished data.). This observation suggested that drug-induced increase of CSH-275 antigen in tumor cells could represent a good strategy for increasing immune response for cancer elimination. We evaluated this hypothesis by testing the ability of Tα1 to modulate tumor antigen expression in an *in vivo* model of metastatic CRC in syngeneic immuno-competent BDIX rats, set up in our laboratories. This preclinical model, in which liver metastases are induced by injection of DHD-K12 cells in the splenic vein, closely mimics the clinical situation of metastatization after surgical resection of primary CRC, with high biological and immunological affinity to the human neoplasia.[42,43] This experimental model has been successfully used in our previous studies in which we assessed the antitumor activity of the triple combination therapy (Tα1 combined with IL-2 and 5 fluorouracil).[27] Using this animal model, we are performing a series of both *in vivo* and *ex vivo* experiments to determine whether Tα1 is capable of modulating the expression of the epitope CSH-275. Results of preliminary experiments are presented. In brief, inbred eight-week-old male BDIX rats were injected with DHD-K12 cells in the splenic vein to induce liver metastases. After 14 days, tumor-injected rats received subcutaneously Tα1

Table 1. Percentage of CSH-275 positive cells in DHD-K12 cells isolated from liver metastases, abdominal mass and ascitic fluid, obtained from treated and untreated BDIX rats

	Liver metastases	Abdominal mass	Ascites
Control	25–32%	45–63%	7–24%
Tα1 200 μg/Kg	32–40%	75–94%	24–32%
Tα1 1 mg/rat	70–95%	73–94%	32–36%

(at a dose of 200 µg/kg or 1 mg/rat) or control diluent, for four days. On day 28, all rats underwent laparotomy, and tumor growth was evaluated by checking nodules on the liver surface, abdominal spread, and the presence of ascites. DHD-K12 cells isolated from dissected metastases and ascites were analyzed for CHS-275 expression by flow cytometry. Results obtained are reported in Figure 1 and Table 1. CSH-275 was expressed in 30% of cultured DHD-K12 cells before intravenous injection and this percentage of CSH-275 positive cells was not significantly modified in liver metastases explanted from untreated rats (range 25–32%). Conversely, cells isolated from abdominal mass showed a higher percentage (45–63%) and cells recovered from ascites a comparatively lower percentage (7–24%) of positive cells. The subcutaneous administration of Tα1 was found capable of inducing a significant increase of CSH-275 expression in all tumor cell subpopulations examined (Table 1 and Fig. 1). Interestingly, in liver metastases, a dose-dependent effect was observed with up to 95% of DHD-K12 cells expressing CSH-275 antigen at the higher Tα1 dose used.

In other preliminary studies, DHD-K12 cells were treated *in vitro* with Tα1 (10 or 50 µg/mL to mimic the circulating concentration occurring in the *in vivo* experiments) and CSH-275 expression was analyzed by confocal microscopy (Fig. 2A) Western blot (Fig. 2B) and flow cytometry (Fig. 2C). Confocal microscopy and cytofluorimetric analyses showed that in DHD-K12 cells Tα1 induced a dramatic and dose-dependent increase of CSH-275 expression in the cytoplasmic compartments, mainly in the Golgi region and endoplasmic reticulum, as well as on cell membrane, as compared to the untreated controls (Fig. 2A and C). Western blot analysis (Fig. 2B) indicated a dose-dependent Tα1-induced increment of CSH-275 expression observed by confocal microscopy. These results suggest that Tα1 treatment stimulates the synthesis of CSH-275 and thereby increases the expression of this antigen on the cell membrane. Therefore, our preliminary data are consistent with the conclusion that Tα1 can upregulate a molecular target relevant for CRC immunotherapy and for generating a CTL response.

Malignant melanoma is increasing in incidence worldwide, and although early primary melanomas can be cured surgically, the metastatic disease has often poor prognosis. Therapeutic options for metastatic disease are limited; however, it has proven to be a disease capable of inducing immune

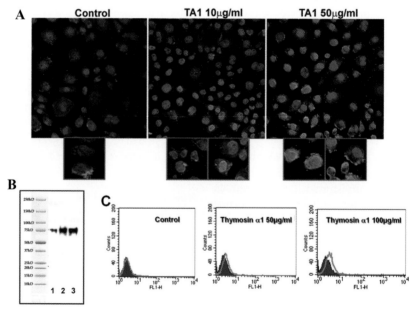

Figure 2. T1 is able to stimulate, in a dose-dependent manner, the expression of the peptide CSH-275 in DHD-K12 cells. (A) Images by confocal microscopy of cells immunostained with the polyclonal antibody specific for CSH (green). (B) Western blot, using the specific polyclonal antibody, on total cell lysates obtained from untreated control cells (lane 1), and cells treated for one hour with increasing doses of thymosin α (lane 2:10µg/mL; lane 3: 50µg/mL). (C) Flow cytometric analysis of Tα1 treatment on CSH-275 expression. These data represent work in progress.

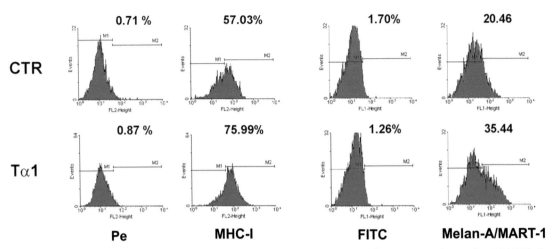

Figure 3. Effect of Tα1 on the expression of MHC-I molecules and of melanoma differentiation antigen Melan-A/MART-1 in TVM-A12 human melanoma cells. Cells were treated with Tα1 10 μg/mL for 48 hours and phenotype modifications were assessed by flow cytometry. For the membrane protein staining, the phycoerythrin (Pe)-conjugated antihuman MHC-I antibody was used. For analysis of the intracellular antigen, cells were permeabilized, stained with mouse antihuman Melan A/MART-1 antibody, and then with FITC-conjugated goat anti-mouse IgG. Samples were analyzed by flow cytometry. Pe, isotype control for MHC-I; FITC, isotype control for Melan A/MART-1. These data represent work in progress.

responses and thus has long been considered a tumor model for the study of cancer immunotherapy. This immunogenicity is often mediated by the recognition of tumor antigens expressed by melanoma cells, which are recognized by CTLs through presentation on HLA molecules.[41]

Different melanoma tumor antigens have been identified.[45] Among these are differentiation antigens, which are only expressed by normal and malignant cells derived from melanocytes but not other cell types[45] and represent potential targets for cancer immunotherapy.[46–48]

The effectiveness of Tα1 in improving the response to chemotherapy to melanoma in animal models as well as in clinical trials has been demonstrated.[26,28–31] To assess whether the antitumor action of Tα1 was mediated not only by its immune stimulatory effect but also by a modulation of antigen expression in tumor cells, we studied the effect of Tα1 on MHC-I molecule and Melan-A/MART-1 antigen expression in a human melanoma cells line (TVM-A12), obtained in our laboratories, from a metastatic melanoma lesion.[49] Melan-A/MART-1 is one of the melanocytic differentiation antigens and considered a potential immunotherapeutic target and a possible diagnostic marker.[41,45–48] It has been already demonstrated by our group that TVM-A12 cells possess peculiar fea-

tures of plasticity, exhibiting an adherent or nonadherent phenotype depending on microenvironmental modifications.[50,51] The transition from the two phenotypes is accompanied by a variety of morphological and molecular alterations. In particular, when grown in serum-free medium, TVM-A12 cells acquire cancer stem cell–like properties, showing reduced levels of the differentiation markers MHC-I and Melan-A/MART-1. Restoration of growth conditions, in the presence of 10% of serum, induces a recovery of MHC-I and Melan-A/MART-1 expression levels, which are significantly increased by the addition of 10 μg/mL Tα1, as compared to untreated control cells (Fig. 3, work in progress). This Tα1-induced effect might result in enhanced immune recognition of melanoma by the cellular effectors of immune system, and reinforces the use of this thymic peptide in cancer therapy.

Conclusions and remarks

Tα1 is an immune-regulatory agent capable of promoting the coordinated activation of the innate and adaptive T helper immunity. It is also capable of directly influencing protein/antigen expression in both normal lymphoid and nonlymphoid cells as well as tumor cells. Its action has been shown to depend on several factors among which are the immature or mature phenotype of cells, the copresence of chemotherapeutic agents and the host immune status, with different responses and outcomes in immunocompetent and immunosuppressed hosts.

It was anticipated that the improvement of knowledge of the molecular targets involved in tumor progression will provide the foundation to define mechanisms of immune escape and novel therapeutic options. Rational therapeutics based on understanding changes in cancer cells and host responses during development will permit rigorous evaluation of cancer treatments and maximize the potential for their success. In this regard, Tα1, because of its dual action on effector and target tumor cells, will require additional studies to demonstrate and define any potential to improve the immune response to tumors.

Acknowledgments

This work was supported by the following grants from the Italian Ministry of University and Research (MIUR): Research Projects of National Interest (PRIN) 2008 to E.G.; MIUR Prot. 10484 to P.S.V. and E.G.; MIUR Prot. 1558 to P.P. We would like to thank Martino T. Miele for administrative support.

Conflicts of interest

The authors declare no conflicts of interest.

References

1. Goldstein, A.L., A. Guha, M.M. Zatz, *et al.* 1972. Purification and biological activity of thymosin, a hormone of the thymus gland. *Proc. Natl. Acad. Sci. U. S. A.* **69:** 1800–1803.
2. Goldstein, A.L. & A.L. Goldstein. 2009. From lab to bedside: emerging clinical applications of thymosin alpha 1. *Expert. Opin. Biol. Ther.* **9:** 593–608.
3. Chretien, P.B., S.D. Lipson, R. Makuch, *et al.* 1978. Thymosin in cancer patients: in vitro effects and correlations with clinical response to thymosin immunotherapy. *Cancer Treat. Rep.* **62:** 1787–1790.
4. Schulof, R.S., M.J. Lloyd, P.A. Cleary, *et al.* 1985. A randomized trial to evaluate the immunorestorative properties of synthetic thymosin-alpha 1 in patients with lung cancer. *J. Biol. Response Mod.* **4:** 147–158.
5. Giuliani, C., G. Napolitano, A. Mastino, *et al.* 2000. Thymosin-alpha1 regulates MHC class I expression in FRTL-5 cells at transcriptional level. *Eur. J. Immunol.* **30:** 778–786.
6. Ahmed, A., D.M. Wong, G.B. Thurman, *et al.* 1979. T-lymphocyte maturation: cell surface markers and immune function induced by T-lymphocyte cell-free products and thymosin polypeptides. *Ann. N.Y. Acad. Sci.* **332:** 81–94.
7. Knutsen, A.P., J.J. Freeman, K.R. Mueller, *et al.* 199 Thymosin-alpha1 stimulates maturation of CD34+ ste cells into CD3+4+ cells in an in vitro thymic epithelia orga coculture model. *Int. J. Immunopharmacol.* **21:** 15–26.
8. Favalli, C., T. Jezzi, A. Mastino, *et al.* 1985. Modulation natural killer activity by thymosin alpha 1 and interfero *Cancer Immunol. Immunother.* **20:** 189–192.
9. Romani, L., F. Bistoni, R. Gaziano, *et al.* 2004. Thymosi alpha 1 activates dendritic cells for antifungal Th1 resi tance through toll-like receptor signaling. *Blood* **103:** 4232 4239.
10. Kabelitz, D., D. Wesch & H.H. Oberg. 2006. Regulation regulatory T cells: role of dendritic cells and toll-like recep tors. *Crit. Rev. Immunol.* **26:** 291–306.
11. Zhang, Y., H. Chen, Y.M. Li, *et al.* 2008. Thymosin al pha 1- and ulinastatin-based immunomodulatory stra egy for sepsis arising from intra-abdominal infection du to carbapenem-resistant bacteria. *J. Infect. Dis.* **198:** 723 730.
12. Thurman, G.B., C. Seals, T.L. Low & A.L. Goldstein. 1984 Restorative effects of thymosin polypeptides on purifie protein derivative–dependent migration inhibition facto production by the peripheral blood lymphocytes of adu thymectomized guinea pigs. *J. Biol. Response Mod.* **3:** 160 173.
13. Ohta, Y., E. Tezuka, S. Tamura & Y. Yagi. 1985. Thymosi alpha 1 exerts protective effect against the 5-FU induce bone marrow toxicity. *Int. J. Immunopharmacol.* **7:** 761 768.
14. Ohta, Y., S. Tamura, S. Nihira, *et al.* 1986. Thymosin alph 1 enhances haemopoietic colony formation by stimulatin the production of interleukin 3 in nu/nu mice. *Int. J. Im munopharmacol.* **8:** 773–779.
15. Kouttab, N.M., A.L. Goldstein, M. Lu, *et al.* 1988. Productior of human B and T cell growth factors is enhanced by thymi hormones. *Immunopharmacology* **16:** 97–105.
16. Huang, K.Y., P.D. Kind, E.M. Jagoda & A.L. Goldstein. 1981 Thymosin treatment modulates production of interferon *J. Interferon. Res.* **1:** 411–420.
17. Shoham, J., M. Cohen, Y. Chandali & A. Avni. 1980. Thymic hormonal activity on human peripheral blood lymphocytes in vitro. I. Reciprocal effect on T and B rosette formation *Immunology* **41:** 353–359.
18. Svedersky, L.P., A. Hui, L. May, *et al.* 1982. Induction and augmentation of mitogen-induced immune interferon production in human peripheral blood lymphocytes by N alpha-desacetylthymosin alpha 1. *Eur. J. Immunol.* **12:** 244–247.
19. Sztein, M.B. & S.A. Serrate. 1989. Characterization of the immunoregulatory properties of thymosin alpha 1 on interleukin-2 production and interleukin-2 receptor expression in normal human lymphocytes. *Int. J. Immunopharma-col.* **11:** 789–800.
20. Sztein, M.B., S.A. Serrate & A.L. Goldstein. 1986. Modulation of interleukin 2 receptor expression on normal human lymphocytes by thymic hormones. *Proc. Natl. Acad. Sci. U. S. A.* **83:** 6107–6111.
21. Leichtling, K.D., S.A. Serrate & M.B. Sztein. 1990. Thymosin alpha 1 modulates the expression of high affinity interleukin-2 receptors on normal human lymphocytes. *Int. J. Immunopharmacol.* **12:** 19–29.

22. Serrate, S.A., R.S. Schulof, L. Leondaridis, *et al.* 1987. Modulation of human natural killer cell cytotoxic activity, lymphokine production, and interleukin 2 receptor expression by thymic hormones. *J. Immunol.* **139:** 2338–2343.

23. Garaci, E., A. Mastino, F. Pica & C. Favalli. 1990. Combination treatment using thymosin alpha 1 and interferon after cyclophosphamide is able to cure Lewis lung carcinoma in mice. *Cancer Immunol. Immunother.* **32:** 154–160.

24. Mastino, A., C. Favalli, S. Grelli, *et al.* 1992. Combination therapy with thymosin alpha 1 potentiates the antitumor activity of interleukin-2 with cyclophosphamide in the treatment of the Lewis lung carcinoma in mice. *Int. J. Cancer* **50:** 493–499.

25. Garaci, E., F. Pica, A. Mastino, *et al.* 1993. Antitumor effect of thymosin alpha 1/interleukin-2 or thymosin alpha 1/interferon alpha,beta following cyclophosphamide in mice injected with highly metastatic Friend erythroleukemia cells. *J. Immunother. Emphasis Tumor Immunol.* **13:** 7–17.

26. Pica, F., M. Fraschetti, C. Matteucci, *et al.* 1998. High doses of Thymosin alpha 1 enhance the antitumor efficacy of combination chemo-immunotherapy for murine B16 melanoma. *Anticancer Res.* **18:** 3571–3578.

27. Rasi, G., G. Silecchia, P. Sinibaldi-Vallebona, *et al.* 1994. Antitumor effect of combined treatment with thymosin alpha 1 and interleukin-2 after 5-fluorouracil in liver metastases from colorectal cancer in rats. *Int. J. Cancer* **57:** 701–705.

28. Garaci, E., M. Lopez, G. Bonsignore, *et al.* 1995. Sequential chemoimmunotherapy for advanced non-small cell lung cancer using cisplatin, etoposide, thymosin alpha 1 and interferon-alpha 2a. *Eur. J. Cancer* **13/14:** 2403–2405.

29. Salvati, F., G. Rasi, L. Portalone, *et al.* 1996. Combined treatment with thymosin alpha 1 and low dose interferon alpha after ifosfamide in non-small cell lung cancer: a phase II controlled trial. *Anticancer Res.* **16:** 1001–1004.

30. Rasi, G., E. Terzoli, F. Izzo, *et al.* 2000. Combined treatment with thymosinalpha1 and low dose interferon-alpha after dacarbazine in advanced melanoma. *Melanoma Res.* **10:** 189–192.

31. Maio, M., A. Mackiewicz, A. Testori, *et al.* 2010. Thymosin Melanoma Investigation Group Large randomized study of thymosin alpha 1, interferon alfa, or both in combination with dacarbazine in patients with metastatic melanoma. *J. Clin. Oncol.* **28:** 1780–1787.

32. Wolf, E., S. Milazzo, K. Boehm, *et al.* 2011. Thymic peptides for treatment of cancer patients. *Cochrane Database Syst Rev.* **16:** 1–61.

33. Shrivastava, P., S.M. Singh & N. Singh. 2004. Effect of thymosin alpha 1 on the antitumor activity of tumor–associated macrophage-derived dendritic cells. *J. Biomed. Sci.* **11:** 623–630.

34. Shrivastava, P., S.M. Singh & N. Singh. 2005. Antitumor activation of peritoneal macrophages by thymosin alpha-1. *Cancer Invest.* **23:** 316–322.

35. Peng, X., P. Zhang, X. Wang, *et al.* 2007. Signaling pathways leading to the activation of IKK and MAPK by thymosin alpha1. *Ann. N.Y. Acad. Sci.* **1112:** 339–350.

36. Pierluigi, B., C. D'Angelo, F. Fallarino, *et al.* 2010. Thymosin alpha1: the regulator of regulators? *Ann. N.Y. Acad. Sci.* **1194:** 1–5.

37. Perruccio, K., P. Bonifazi, F. Topini, *et al.* 2010. Thymosin alpha1 to harness immunity to pathogens after haploidentical hematopoietic transplantation. *Ann. N.Y. Acad. Sci.* **1194:** 153–161.

38. Matteucci, C., A. Minutolo, P. Sinibaldi-Vallebona, *et al.* 2010. Transcription profile of human lymphocytes following in vitro treatment with thymosin alpha-1. *Ann N Y Acad Sci.* **1194:** 6–19.

39. Van den Eynde, B.J. & P. van der Bruggen. 1997. T cell defined tumor antigens. *Curr. Opin. Immunol.* **9:** 684–693.

40. Boon, T. & L.J. Old. 1997. Cancer tumor antigens. *Curr. Opin. Immunol.* **9:** 681–683.

41. Garaci, E., C. Favalli, F. Pica, *et al.* 2007. Thymosin alpha 1: from bench to bedside. *Ann. N.Y. Acad. Sci.* **1112:** 225–234.

42. Guadagni, F., P. Graziano, M. Roselli, *et al.* 1999. Differential expression of a new tumor-associated antigen,TLP, during human colorectal cancer tumorigenesis. *Am. J. Pathol.* **154:** 993–999.

43. Rasi, G., P. Sinibaldi-Vallebona, A. Serafino, *et al.* 2000. A new human tumor-associated antigen (TLP) is naturally expressed in rat DHD-K12 colorectal tumor cells. *Int. J. Cancer.* **85:** 540–544.

44. Serafino, A., N. Moroni, R. Psaila, *et al.* 2012. Antiproliferative effect of atrial natriuretic peptide on colorectal cancer cells: evidence for an Akt-mediated cross-talk between NHE-1 activity and Wnt/β-catenin signaling. *Biochim. Biophys. Acta* **1822:** 1004–1018.

45. Barrow, C., J. Browning, D. MacGregor, *et al.* 2006. Tumor antigen expression in melanoma varies according to antigen and stage. *Clin. Cancer Res.* **12:** 764–771.

46. Kawakami, Y. & S.A. Rosenberg. 1997. Human tumor antigens recognized by T cells. *Immunol. Res.* **16:** 313–339.

47. Davis, I.D., M. Jefford, P. Parente & J. Cebon. 2003. Rational approaches to human cancer immunotherapy. *J. Leukoc. Biol.* **73:** 3–29.

48. Rosenberg, S.A. 2004. Shedding light on immunotherapy for cancer. *N. Engl. J. Med.* **350:** 1461–1463.

49. Melino, G., P. Sinibaldi-Vallebona, S. D'Atri, *et al.* 1993. Characterization of three melanoma cell lines (TVM-A12, TVM-A197, TVM-BO) sensitivity to lysis and effect of retinoic acid. *Clin. Chem. Enzym. Comms.* **6:** 105–119.

50. Serafino, A., E. Balestrieri, P. Pierimarchi, *et al.* 2009. The activation of human endogenous retrovirus K (HERV-K) is implicated in melanoma cell malignant transformation. *Exp. Cell Res.* **315:** 849–862.

51. Sinibaldi Vallebona, P., C. Matteucci & C. Spadafora. 2011. Retrotransposon-encoded reverse transcriptase in the genesis, progression and cellular plasticity of human cancer. *Cancers* **3:** 1141–1115.

Ann. N.Y. Acad. Sci. ISSN 0077-8923

ANNALS OF THE NEW YORK ACADEMY OF SCIENCES

Issue: *Thymosins in Health and Disease*

Prothymosin α plays multifunctional cell robustness roles in genomic, epigenetic, and nongenomic mechanisms

Hiroshi Ueda, Hayato Matsunaga, and Sebok Kumar Halder

Department of Molecular Pharmacology and Neuroscience, Nagasaki University Graduate School of Biomedical Sciences, Bunkyo-machi, Nagasaki, Japan

Address for correspondence: Hiroshi Ueda, Department of Molecular Pharmacology and Neuroscience, Nagasaki University Graduate School of Biomedical Sciences, 1-14 Bunkyo-machi, Nagasaki, 852-8521, Japan. ueda@nagasaki-u.ac.jp

Prothymosin α (ProTα) possesses multiple functions for cell robustness. This protein functions intracellularly to stimulate cell proliferation and differentiation through epigenetic or genomic mechanisms. ProTα also regulates the cell defensive mechanisms through an interaction with the Nrf2-Keap1 system. Under the apoptotic conditions, it inhibits apoptosome formation by binding to Apaf-1. Regarding extracellular functions, ProTα is extracellularly released from the nucleus upon necrosis-inducing ischemia stress in a manner of nonclassical release, and thereby inhibits necrosis. However, under the condition of apoptosis, the C-terminus of ProTα is cleaved off and loses binding activity to cargo protein S100A13 for nonclassical release. However, cleaved ProTα is retained in the cytosol and inhibits apoptosome formation. ProTα was recently reported to cause immunological actions through the Toll-like receptor 4. However, the authors also suggest the possible existence of additional receptors for robust cell activities against ischemia stress.

Keywords: linker histone H1; CBP/p300; estrogen receptor; Nrf2-Keap1; Apaf-1; Toll-like receptor-4

Introduction

ProTα is a highly acidic nuclear protein of the α-thymosin family and is found in the nuclei of virtually all mammalian cells.[1,2] ProTα is generally thought to be an oncoprotein that is correlated with cell proliferation by sequestering anticoactivator factor, a repressor of estrogen receptor activity, in various cells.[3,4] With regard to cell death regulation, intracellular ProTα was reported to play a cytoprotective role by inhibiting apoptosome formation in HeLa cells subjected to apoptotic stress.[5] On the other hand, ProTα has been reported to act as an extracellular signaling molecule, as observed in the activation of macrophages, natural killer cells, and lymphokine-activated killer cells, and in the production of IL-2 and TNF-α.[6] ProTα was most recently reported to exert immune responses through Toll-like receptor 4.[7,8] It should be noted that ProTα has potent neuroprotective actions through unique mechanisms by inhibiting neuronal necrosis.[9,10] A recent study revealed the machinery of nonclassical/nonvesicular release of ProTα from the nuclei upon the ischemia/cell starving stress.[11] Thus, there are accumulating findings supporting the conclusion that ProTα plays multiple roles inside and outside of the cell, particularly for cell survival and proliferation.[12]

Nuclear functions

Epigenetic mechanisms

ProTα is highly acidic (pI = 3.55) owing to its abundance of glutamic and aspartic acids (approximately 50% of the total amino acid residues) in the middle part of the protein. The cluster of acidic amino acids in this region seems to resemble a putative histone-binding domain, being consistent with the fact that there is a nuclear localization signal (NLS) at the C-terminal end (human ProTα: KR and KKQK at 87 and 101, respectively). Indeed there are reports that ProTα is highly expressed in many different types of cancer cells,[13–20] and closely related to the cell proliferation and differentiation.[13,21–23] Nuclear ProTα epigenetically stimulates

doi: 10.1111/j.1749-6632.2012.06675.x

Ann. N.Y. Acad. Sci. 1269 (2012) 34–43 © 2012 New York Academy of Sciences.

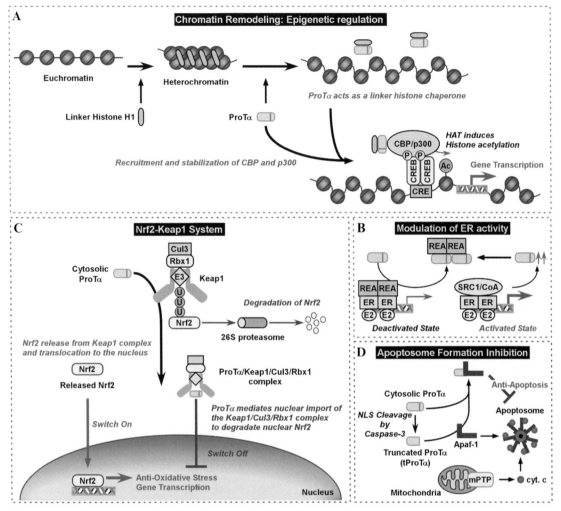

Figure 1. Intracellular multiple functions of ProTα for cell survival and proliferation through a variety of protein–protein interaction. (A) ProTα mediates chromatin remodeling and gene transcription. ProTα acts as a linker histone chaperone. Histone H1 binds to nucleosomal DNA and induces chromatin condensation. ProTα can facilitate H1 displacement from and deposition onto the chromatin template. ProTα is involved in histone acetylation by recruitment and stabilization of p300 histone acetyl transferase (HAT) and CREB-binding protein (CBP). (B) Enhancement of estrogen receptor (ER) transcriptional activity by binding to its repressor (REA: repressor estrogen receptor activity). (C) ProTα regulates the Nrf2-Keap1 system. ProTα binds to Keap1-releasing transcription factor Nrf2, which in turn upregulates antioxidative stress genes involving in apoptosis and autophagy (depicted as switch on). Meanwhile ProTα mediates nuclear import of Keap1/Cul3/Rbx1 complex leads to ubiquitination and degradation of Nrf2 (switch off). (D) ProTα -mediated inhibition of apoptosome by binding of to Apaf-1.

gene transcription by binding to histones,[12] p300 histone acetyltransferase,[24] and CREB-binding protein (CBP).[25] These findings suggests the role of ProTα in the chromatin remodeling (Fig. 1A). According to this hypothesis, ProTα detaches linker histone H1 from chromatin[26] and thus "loosens" it. ProTα also recruits CBP/p300 and stabilizes CBP/p300-CREB complex, resulting in the enhancement of CRE-regulated gene transcrip-

tion through histone acetylation and chromatin remodeling.

Modulation of estrogen receptor (ER)–mediated transcriptional activity

ProTα enhances the transcriptional activity of estrogen (E2) receptor through a removal of a repressor (estrogen receptor activity (REA)/B cell receptor-associated protein BAP37/prohibitin-2).[3,27] In the presence of E2, REA, and coactivator SRC-1 compete for the binding to the ER. ProTα causes a dissociation of REA from ER and shifts the ER status to the SRC-1-binding, or transcription-active, form (Fig. 1B).

Cytosol functions

Switch on/off of Nrf2-Keap1 mechanisms
ProTα possessing an NLS at the C-terminus is localized in the nucleus in most cells. The nuclear import of these molecules with an NLS is mediated by importin α. As the nucleus-to-cytosol export of ProTα is presumed to occur in a passive diffusion due to its smaller size,[11] it is natural that ProTα may have some biological actions in the cytosol, as well as in the nucleus. The Nrf2 (nuclear factor erythroid 2-related factor 2)-Keap1 (Kelch-like ECH-associated protein 1) system is known to play roles in the cell adaptation to oxidative and electrophilic stress.[28–30] Nrf2 is a nuclear transcription factor that regulates expression of several defensive genes, including detoxifying enzymes, and antioxidant genes.[28–30] In the absence of stress, cytosol Nrf2 is "trapped" by Keap1, ubiquitinated by the Cul3/Rbx1-dependent E3 ubiquitin ligase, and then subsequently degraded by the 26S proteasome.[31] Recent studies revealed that ProTα binds to Keap 1 and releases Nrf2 from the Nrf2-Keap1 complex ("switch on").[32] Interestingly, ProTα also mediates nuclear import of the Keap1/Cul3/Rbx1 complex to degrade nuclear Nrf2 ("switch off").[33] Thus, it is speculated that ProTα regulates the cell-defensive roles of Nrf2 by switch on/off mechanisms (Fig. 1C).

Inhibition of apoptosome formation
When the cell is under mitochondrial stress, such as in the case with a growth factor deprivation, cytochrome c (cyt. c)—a member of soluble mitochondrial intermembrane proteins (SIMPs)—is released from mitochondria (Fig. 2A). As a result, the apoptosome, composed of cyt. c and Apaf-1, is formed and followed by a cascade of caspase activation and apoptotic DNA fragmentation.[34,35] Recent studies revealed that ProTα inhibits apoptosome formation by interaction with Apaf-1.[5,36] During apoptosis, on the other hand, ProTα loses its C-terminus containing the NLS, due to the digestion by caspase-3, and is redistributed to the cytosol.[11,37,38] As both cytosolic full-length ProTα and truncated ProTα have a similar potential of apoptosome inhi-

bition, it is interesting to speculate that ProTα has a potential as a natural inhibitor of apoptosis (Fig. 1D). This speculation is further supported by a report that ProTα inhibits apoptogenic compound–induced apoptosis by interacting with p8 (a nuclear protein-1 and candidate of metastasis-1).[39,40]

Extracellular functions

Identification of ProTα as a neuronal necrosis inhibitory factor
Based on the findings that cortical neurons at a low density rapidly die by necrosis under a serum-free or starving condition and that survival is density dependent,[10] we previously attempted to search for antinecrotic factors from the conditioned medium by using molecular weight cut-off ultrafiltration, ion-exchange filtration, SDS-PAGE separation, and matrix-assisted laser desorption/ionization-time of flight mass spectrometry (MALDI-TOF MS). A subsequent search of the nonredundant NCBI protein database for matching peptide mass fingerprints revealed 17 peptides consistent with the conclusion that the only active substance was acetylated ProTα.[9] After various approaches, we discovered an efficient way to obtain significant amounts of active materials that were unique to rat ProTα. Moreover, tandem MS analysis confirmed that the N-terminal of purified ProTα was an acetylated serine, in agreement with a previous report.[9]

Non-classical release mechanisms
ProTα was discovered in conditioned medium from serum-free or starving stress conditions of cultured cortical neurons. To examine the molecular basis of stress-induced ProTα release, we used C6 glioma (astrocytoma) cells because of their robustness against ischemia stress. On the analogy of the nonclassical release of FGF-1,[41,42] which lacks signal peptide sequence, ischemic stress caused a limiting extracellular release of ProTα (which also lacks signal peptide sequence). Our study revealed that the mechanisms underlying the nonclassical ProTα release from C6 glioma cells are mediated by the loss of ATP and Ca^{2+} influx through N-type voltage-dependent Ca^{2+} channel activity.[11] In this mechanism (Fig. 3A), the first step is the release of ProTα from the nucleus (passive diffusion) due to ATP loss, followed by a Ca^{2+}-dependent interaction with Ca^{2+} binding protein S100A13, a cargo protein. The release of ProTα from the nucleus upon an ischemic

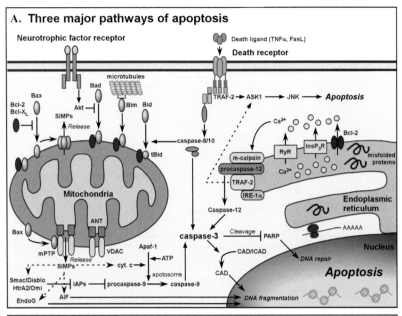

A. Three major pathways of apoptosis

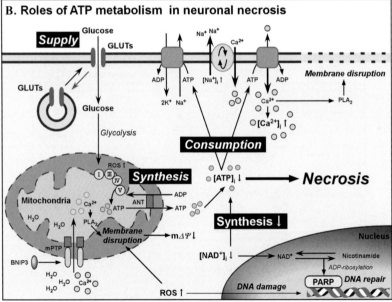

B. Roles of ATP metabolism in neuronal necrosis

Figure 2. Schematic illustration of apoptosis and necrosis. (A) Three major pathways of apoptosis. Mitochondrial pathways are closely related to the expression of members of the Bcl-2 family of proteins. Proapoptotic Bak and Bax open mitochondrial permeability transition pores (mPTPs) to release soluble intermembrane proteins (SIMPs), including cytochrome c (cyt.c), apoptosis-inducing factor (AIF), Smac/DIABLO, EndoG, and HtrA2/Omi. Among these, cyt.c plays a major role in inducing apoptosis through activation of caspase-3 and caspase-activated DNase (CAD). Bcl-2 and Bcl-XL are major antiapoptotic proteins that inhibit the functions of these proapoptotic proteins. The other two pathways through death receptors (FAS and TNF-α receptors) or endoplasmic reticulum stress also use caspase-3 activation as the common execution pathway. (B) Roles of ATP metabolism in neuronal necrosis. Glucose transporters (GLUTs) are involved in the supply of cellular glucose (depicted as supply), a substrate for ATP production through glycolysis and oxidative phosphorylation (depicted as synthesis) in mitochondria. Some species of GLUT are constitutively localized, while others are translocated to the membrane upon cell stimulation by extracellular signals. Abundant cellular ATP molecules maintain intracellular ionic behavior (depicted as consumption). Poly (ADP-ribose) polymerase (PARP) restores the DNA damage caused by cellular stress, by using abundant NAD$^+$ molecules. NAD$^+$ reduction induces the decline of NADH-dependent ATP synthesis. A rapid decrease in the cellular ATP levels leads to necrosis.

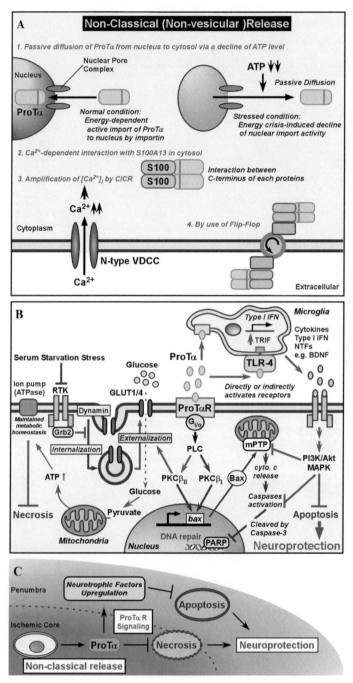

Figure 3. Schematic illustration for the extracellular release of ProTα and its neuroprotective roles in ischemic condition. (A) Necrotic stress–induced nonclassical release of ProTα. Necrosis-dependent energy crisis induces decline of nuclear import activity, subsequently ProTα, a small protein is passively diffused from the nucleus to the cytosol. Interaction between ProTα and extacellular cargo protein S100A13 is a Ca^{2+}-dependent manner through N-type voltage-dependent Ca^{2+} channel (VDCC) activity. Intracellular $[Ca^{2+}]_i$ is amplified by the mechanism of Ca^{2+}-induced Ca^{2+} release (CICR). Extracellular release of ProTα-S100A13 complex is driven by use of a flip-flop mechanism. (B) Mechanism of ProTα-induced cell death mode switch and neuroprotection. Serum-free or starvation stress leads to endocytosis of the glucose transporters GLUT1/4, which in turn causes bioenergetic catastrophe-mediated necrosis through a rapid loss of glucose supply. Addition of ProTα to ischemia-treated neurons causes translocation of GLUT1/4 to the membrane to allow sufficient glucose supply through activation of $G_{i/o}$, PLC and $PKC\beta_{II}$. ProTα-induced apoptosis occurs

condition, causing rapid decrease in cellular ATP levels, is explained by the following possible mechanism. To retain ProTα in the nucleus against passive diffusion, the repeated recycling of importin α between cytosol and nucleus is required. Ran, a small G protein, is known to execute this importin recycling process.[43,44] Therefore, the energy crisis—or cellular ATP loss—leads to a decrease in GTP levels and disables importin recycling due to the decreased level of Ran in an active GTP-binding form.

It should be noted that there was no ProTα release when the cells were treated with apoptogenic reagents, though ProTα is released to the cytosol from the nucleus.[11] Detailed studies revealed that caspase-3, activated by apoptogenic reagent treatments, cleaves the C-terminus of ProTα, which contains an NLS and a key domain responsible for the interaction with S100A13. These findings enable us to speculate that cytosolic ProTα is redistributed to the cytosol from the nucleus under apoptotic conditions and may have an antiapoptotic self-defensive function, as mentioned above.

Antinecrosis mechanisms

Under the serum-free condition without any supplements, neurons rapidly die by necrosis, as seen in the typical necrosis features, such as disrupted plasma membranes and swollen mitochondria (in TEM analysis), and rapid decreases in $[^3H]$-2-deoxyglucose uptake and related cellular ATP levels.[9] Pharmacological studies revealed that the survival activity of recombinant ProTα is mediated through activation of phospholipase C (PLC) and protein kinase C (PKC) β.[9] Quite similar changes were reproduced in the low-oxygen and low-glucose (LOG) ischemia-reperfusion culture model. Addi-

tion of ProTα reversed the rapid decrease in the cellular ATP levels of cortical neurons following LOG-ischemic stress and reperfusion with serum-containing medium.[9] We previously reported that the membrane translocation of the glucose transporters GLUT1/4 is largely inhibited in serum-free cultures of cortical neurons, which leads to necrotic cell death.[45,46] In fact, LOG-stress decreased GLUT1/4 membrane translocation, as evaluated by immunocytochemistry and Western blot analysis, and addition of ProTα reversed these changes. This cell biological change with ProTα-induced GLUT1/4 membrane translocation enabled us to successfully characterize the putative receptor signaling mechanisms.[9] The ProTα-induced GLUT1/4 translocation was blocked by the treatments with pertussis toxin and PLC and PKCβ inhibitors. More specifically, the action was abolished by PKCβ$_{II}$ antisense oligodeoxynucleotide (AS-ODN), though ProTα activates PKC isoforms, α, β$_I$, and β$_{II}$, but not γ, δ, ε, or ς. Taken together, the pharmacological studies revealed that the ProTα-induced membrane translocation of GLUT1/4, which underlies the mechanisms for necrosis inhibition, is mediated through activation of putative G$_{i/o}$-coupled receptor, PLC and PKCβ$_{II}$ (Fig. 3B).

Inhibition of rapid necrosis by caspase activation

Although the addition of ProTα delayed the cell death of cortical neurons in serum-free culture, most of the neurons completely died by apoptosis after 24 hours. However, when neurons were treated with ProTα under conditions of ischemia and subsequent reperfusion with serum-containing medium, no significant cell death was observed

---

later at 12 h after the start of serum-free stress. The machinery is mediated by upregulation of Bax, which in turn causes mitochondrial cyto c release and subsequent apoptosis. Bax upregulation is also mediated by activation of G$_{i/o}$, PLC, and PKC, similar to the case for necrosis. However, both PKCβ$_I$ and PKCβ$_{II}$ upregulations mediate this apoptotic mechanism. Since caspase-3–mediated PARP degradation minimizes the ATP consumption, the apoptosis induction may have a crucial role in inhibiting the rapid necrosis. In addition, since pyruvate, a substrate for ATP production in mitochondria, inhibits necrosis but does not cause apoptosis, the apoptosis machinery seems to be independent of the necrosis inhibition. Neurotrophins, such as BDNF or EPO and type I IFN, which are expected in the ischemic brain and retina, can inhibit the apoptosis machinery at a later stage. (C) *In vivo* neuroprotective role of ProTα in the ischemic brain and retina. ProTα is first released upon ischemic stress in the ischemic core. Released ProTα exerts a suppression of the necrosis of neighboring neurons, which play a role of the early stage of neuroprotection. Expression of neurotrophic factors, such as BDNF and EPO will then occur and block the apoptosis in the penumbra (late stage of neuroprotection). Although dead cell-derived cytotoxic molecules also cause late apoptosis in the penumbra, the initial blockade of necrosis may minimize the occurrence of late apoptosis. This view may be consistent with the rationale for therapies for acute ischemic stroke.

---

for at least 48 hours.[9] These findings indicate that serum factors prevented ProTα-induced apoptosis. Indeed, further addition of nerve growth factor (NGF), brain-derived growth factor (BDNF), basic fibroblast growth factor (bFGF), or interleukin (IL)-6—representative apoptosis inhibitors[47–51]—rescued the cell survival in serum-free culture for 48 h, while these factors alone had no effects on the survival.[9] Similar effects were observed with BIP-V5, which blocks the translocation of Bax to mitochondria.[52] BIP-V5 also selectively inhibited the ProTα-induced apoptosis.

However, mixed addition of ProTα and N-benzyloxycarbonyl-Val-Ala-Asp (OMe)-fluoromethylketone (zVAD-fmk), a pan-type caspase inhibitor, caused necrosis at the later stage.[9,10] This unexpected results may be explained by the view that ProTα-induced continuous cyto.c release causes a loss of ATP and necrosis induction. The reason why ProTα alone causes apoptosis without causing necrosis may be explained by the fact that caspase-3 cleaves poly-(ADP-ribose) polymerase, PARP, which heavily uses NAD, and in turn leads to a decreased ATP synthesis (Fig. 2B). In other words, apoptosis-induction in the early stage after ischemia may play a defensive role in inhibiting rapid cell death by necrosis (Fig. 3C).

In vivo neuroprotection

Systemic and local injections of ProTα markedly inhibits the histological and functional damage induced by cerebral and retinal ischemia.[53,54] Although ProTα inhibits apoptosis as well as necrosis in *in vivo* studies, the discrepancy from *in vitro* studies may be explained by the fact that antiapoptotic factors may inhibited the apoptotic machineries induced by ProTα. This speculation was successfully tested by the *in vivo* administration of antibodies against either brain-derived neurotrophic factor (BDNF) or erythropoietin (EPO), which restores apoptosis in cerebral and retinal ischemia models. Although ProTα upregulates BDNF levels in the retina only in the presence of ischemia, the underlying mechanisms remain elusive. Most recently, it was reported that ProTα causes immune responses through Toll-like receptor 4 (TLR4).[7,8] In this study, exogenous full-length ProTα, and endogenous ProTα released by CD8[+] T cells, may act as a signaling ligand for TLR4 and trigger the TRIF-mediated IFN-β induction and MyD88-

mediated induction of proinflammatory cytokines, such as TNF-α, to suppress HIV-1 after the entry into macrophages.[7] As there are reports that stroke-induced brain damage is inhibited by a preconditioning treatment with LPS (an activator of TLR4),[55,56] this mechanism might cause the *in vivo* induction of antiapoptotic factors. There are several reports that TLRs cause induction of type-I IFN, which can be neroprotective.[56,57] TLR-4–mediated TRIF-IRF3 signaling may underlie major mechanisms for neuroprotection against stroke.[56,58] It is of interest that heterotrimeric G protein $G_{i/o}$ is important for the activation of MAPK and Akt downstream of TLR4, as well as for the full activation of IFN signaling downstream of TRIF-dependent signaling via TLR4.[59–61] The involvement of $G_{i/o}$ would be also consistent with the ProTα signaling, in terms of the antinecrosis pathway (see Fig. 3B).[9] In our preliminary studies, however, the preconditioning treatment of ProTα only partially inhibited retinal ischemia damage, while posttreatment with ProTα 24 h after the ischemia completely inhibited damage. Therefore, it is also speculated that additional receptor systems for ProTα play a more important for the retinal protection system.

Conclusions

ProTα and thymosin α1 possess a number of extracellular cytokine–like functions, including stimulating up-regulation of HA-DR, IL-2 receptor, dendritic cell maturation, chemotaxis, and possible antiviral, anticancer, and antifungal activities.[62–65] As we do not detect any similar neuroprotective actions with thymosin α1 for necrosis inhibition and neuroprotective actions, it is evident that the machineries underlying beneficial activities of both biologically active molecules are different. The most important take-home messages here is that ProTα plays key roles in the survival activity of intact cells, inhibits necrosis under the condition of neuronal necrosis, and inhibits apoptosis under the condition of apoptosis. The hypothesis is that ProTα acts as a "robustness" or cell death mode switch molecule from uncontrollable necrosis to neurotrophin-reversible apoptosis, and may provide a promising novel strategy for preventing serious damage in stroke (Fig. 3C). However, clarification of mechanisms underlying the intrinsic robust activity of ProTα must wait for the identification of ProTα-binding proteins, including the $G_{i/o}$-coupled receptor for cell death mode switch. Further studies to examine how these candidate molecules play such multiple functions in cell death regulation are the next exciting step.

Acknowledgments

Parts of this study were supported by Grants-in-Aid for Scientific Research (to H.U., B: 13470490 and B: 15390028), for Young Scientists (to H.M., B: 20770105), for Exploratory Research (to H.M., 22657306), on Priority Areas—Research on Pathomechanisms of Brain Disorders (to H.U., 17025031, 18023028, 20023022) from the Ministry of Education, Culture, Sports, Science and Technology (MEXT), the Japan Society for the Promotion of Science (JSPS); and Health and Labour Sciences Research Grants on Research on Biological Resources and Animal Models for Drug Development (to H.U., H20-Research on Biological Resources and Animal Models for Drug Development-003) from the Ministry of Health, Labour and Welfare.

Conflicts of interest

The authors declare no conflicts of interest.

References

1. Haritos, A.A., J. Caldarella & B.L. Horecker. 1985. Simultaneous isolation and determination of prothymosin alpha, parathymosin alpha, thymosin beta 4, and thymosin beta 10. *Anal. Biochem.* **144:** 436–440.
2. Clinton, M., L. Graeve, H. el-Dorry, *et al.* 1991. Evidence for nuclear targeting of prothymosin and parathymosin synthesized in situ. *Proc. Natl. Acad. Sci. USA* **88:** 6608–6612.
3. Martini, P.G., R. Delage-Mourroux, D.M. Kraichely & B.S. Katzenellenbogen. 2000. Prothymosin alpha selectively enhances estrogen receptor transcriptional activity by interacting with a repressor of estrogen receptor activity. *Mol. Cell Biol.* **20:** 6224–6232.
4. Bianco, N.R. & M.M. Montano. 2002. Regulation of prothymosin alpha by estrogen receptor alpha: molecular mechanisms and relevance in estrogen-mediated breast cell growth. *Oncogene* **21:** 5233–5244.
5. Jiang, X., H.E. Kim, H. Shu, *et al.* 2003. Distinctive roles of PHAP proteins and prothymosin-alpha in a death regulatory pathway. *Science* **299:** 223–226.
6. Piñeiro, A., O.J. Cordero & M. Nogueira. 2000. Fifteen years of prothymosin alpha: contradictory past and new horizons. *Peptides* **21:** 1433–1446.
7. Mosoian, A., A. Teixeira, C.S. Burns, *et al.* 2010. Prothymosin-alpha inhibits HIV-1 via Toll-like receptor 4-mediated type I interferon induction. *Proc. Natl. Acad. Sci. USA* **107:** 10178–10183.
8. Mosoian, A. 2011. Intracellular and extracellular cytokine-like functions of prothymosin α: implications for the development of immunotherapies. *Future Med. Chem.* **3:** 1199–1208.
9. Ueda, H., R. Fujita, A. Yoshida, *et al.* 2007. Identification of prothymosin-alpha1, the necrosis-apoptosis switch molecule in cortical neuronal cultures. *J. Cell Biol.* **176:** 853–862.
10. Ueda, H. 2009. Prothymosin alpha and cell death mode switch, a novel target for the prevention of cerebral ischemia-induced damage. *Pharmacol. Ther.* **123:** 323–333.
11. Matsunaga, H. & H. Ueda. 2010. Stress-induced nonvesicular release of prothymosin-á initiated by an interaction with S100A13, and its blockade by caspase-3 cleavage. *Cell Death Differ* **17:** 1760–1772.
12. Gómez-Márquez, J. 2007. Function of prothymosin alpha in chromatin decondensation and expression of thymosin beta-4 linked to angiogenesis and synaptic plasticity. *Ann. N.Y. Acad. Sci.* **1112:** 201–209.
13. Dominguez, F., C. Magdalena, E. Cancio, *et al.* 1993. Tissue concentrations of prothymosin alpha: a novel proliferation index of primary breast cancer. *Eur. J. Cancer* **29A:** 893–897.
14. Wu, C.G., N.A. Habib, R.R. Mitry, *et al.* 1997. Overexpression of hepatic prothymosin alpha, a novel marker for human hepatocellular carcinoma. *Br. J. Cancer* **76:** 1199–1204.
15. Magdalena, C., F. Dominguez, L. Loidi & J.L. Puente. 2000. Tumour prothymosin alpha content, a potential prognostic marker for primary breast cancer. *Br. J. Cancer* **82:** 584–590.
16. Sasaki, H., Y. Sato, S. Kondo, *et al.* 2001. Expression of the prothymosin alpha mRNA correlated with that of N-myc in neuroblastoma. *Cancer Lett.* **168:** 191–195.
17. Hapke, S., H. Kessler, B. Luber, *et al.* 2003. Ovarian cancer cell proliferation and motility is induced by engagement of integrin alpha(v)beta3/Vitronectin interaction. *Biol. Chem.* **384:** 1073–1083.
18. Shiwa, M., Y. Nishimura, R. Wakatabe, *et al.* 2003. Rapid discovery and identification of a tissue-specific tumor biomarker from 39 human cancer cell lines using the SELDI ProteinChip platform. *Biochem. Biophys. Res. Commun.* **309:** 18–25.
19. Leys, C.M., S. Nomura, B.J. LaFleur, *et al.* 2007. Expression and prognostic significance of prothymosin-alpha and ERp57 in human gastric cancer. *Surgery* **141:** 41–50.
20. Kashat, L., A. So, O. Masui, *et al.* 2010. Secretome based Identification and Characterization of Potential Biomarkers in Thyroid Cancer. *J. Proteome Res.* **9:** 5757–5769.
21. Gómez-Márquez, J., F. Segade, M. Dosil, *et al.* 1989. The expression of prothymosin alpha gene in T lymphocytes and leukemic lymphoid cells is tied to lymphocyte proliferation. *J. Biol. Chem.* **264:** 8451–8454.
22. Conteas, C.N., M.G. Mutchnick, K.C. Palmer, *et al.* 1990. Cellular levels of thymosin immunoreactive peptides are linked to proliferative events: evidence for a nuclear site of action. *Proc. Natl. Acad. Sci. USA* **87:** 3269–3273.
23. Smith, M.R., A. al-Katib, R. Mohammad, *et al.* 1993. Prothymosin alpha gene expression correlates with proliferation, not differentiation, of HL-60 cells. *Blood* **82:** 1127–1132.
24. Subramanian, C., S. Hasan, M. Rowe, *et al.* 2002. Epstein–Barr virus nuclear antigen 3C and prothymosin alpha interact with the p300 transcriptional coactivator at the CH1 and CH3/HAT domains and cooperate in regulation of transcription and histone acetylation. *J. Virol.* **76:** 4699–4708.

25. Karetsou, Z., A. Kretsovali, C. Murphy, *et al.* 2002. Prothymosin alpha interacts with the CREB-binding protein and potentiates transcription. *EMBO Rep.* **3:** 361–366.

26. Happel, N. & D. Doenecke. 2009. Histone H1 and its isoforms: contribution to chromatin structure and function. *Gene* **431:** 1–12.

27. Martini, P.G. & B.S. Katzenellenbogen. 2003. Modulation of estrogen receptor activity by selective coregulators. *J. Steroid Biochem. Mol. Biol.* **85:** 117–122.

28. Dhakshinamoorthy, S., D.J. Long & A.K. Jaiswal. 2000. Antioxidant regulation of genes encoding enzymes that detoxify xenobiotics and carcinogens. *Curr. Top Cell Regul.* **36:** 201–216.

29. Jaiswal, A.K. 2000. Regulation of genes encoding NAD(P)H:quinone oxidoreductases. *Free Radic. Biol. Med.* **29:** 254–262.

30. Kobayashi, M. & M. Yamamoto. 2006. Nrf2-Keap1 regulation of cellular defense mechanisms against electrophiles and reactive oxygen species. *Adv. Enzyme Regul.* **46:** 113–140.

31. McMahon, M., N. Thomas, K. Itoh, *et al.* 2006. Dimerization of substrate adaptors can facilitate cullin-mediated ubiquitylation of proteins by a "tethering" mechanism: a two-site interaction model for the Nrf2-Keap1 complex. *J. Biol. Chem.* **281:** 24756–24768.

32. Karapetian, R.N., A.G. Evstafieva, I.S. Abaeva, *et al.* 2005. Nuclear oncoprotein prothymosin alpha is a partner of Keap1: implications for expression of oxidative stress-protecting genes. *Mol. Cell Biol.* **25:** 1089–1099.

33. Niture, S.K. & A.K. Jaiswal. 2009. Prothymosin-alpha mediates nuclear import of the INrf2/Cul3 Rbx1 complex to degrade nuclear Nrf2. *J. Biol. Chem.* **284:** 13856–13868.

34. Schafer, Z.T. & S. Kornbluth. 2006. The apoptosome: physiological, developmental, and pathological modes of regulation. *Dev. Cell* **10:** 549–561.

35. Reubold, T.F. & S. Eschenburg. 2012. A molecular view on signal transduction by the apoptosome. *Cell Signal* **24:** 1420–1425.

36. Qi, X., L. Wang & F. Du. 2010. Novel Small Molecules Relieve Prothymosin alpha-Mediated Inhibition of Apoptosome Formation by Blocking Its Interaction with Apaf-1. *Biochemistry* **49:** 1923–1930.

37. Evstafieva, A.G., G.A. Belov, M. Kalkum, *et al.* 2000. Prothymosin alpha fragmentation in apoptosis. *FEBS Lett.* **467:** 150–154.

38. Evstafieva, A.G., G.A. Belov, Y.P. Rubtsov, *et al.* 2003. Apoptosis-related fragmentation, translocation, and properties of human prothymosin alpha. *Exp. Cell Res.* **284:** 211–223.

39. Malicet, C., J.C. Dagorn, J.L. Neira & J.L. Iovanna. 2006. p8 and prothymosin alpha: unity is strength. *Cell Cycle* **5:** 829–830.

40. Malicet, C., V. Giroux, S. Vasseur, *et al.* 2006. Regulation of apoptosis by the p8/prothymosin alpha complex. *Proc. Natl. Acad. Sci. USA* **103:** 2671–2676.

41. Matsunaga, H. & H. Ueda. 2006. Voltage-dependent N-type Ca^{2+} channel activity regulates the interaction between FGF-1 and S100A13 for stress-induced non-vesicular release. *Cell Mol. Neurobiol.* **26:** 237–246.

42. Matsunaga, H. & H. Ueda. 2006. Evidence for serum-deprivation-induced co-release of FGF-1 and S100A13 from astrocytes. *Neurochem. Int.* **49:** 294–303.

43. Yasuda, Y., Y. Miyamoto, T. Saiwaki & Y. Yoneda. 2006. Mechanism of the stress-induced collapse of the Ran distribution. *Exp. Cell Res.* **312:** 512–520.

44. Yasuhara, N., M. Oka & Y. Yoneda. 2009. The role of the nuclear transport system in cell differentiation. *Semin Cell Dev. Biol.* **20:** 590–599.

45. Fujita, R. & H. Ueda. 2003. Protein kinase C-mediated necrosis-apoptosis switch of cortical neurons by conditioned medium factors secreted under the serum-free stress. *Cell Death Differ* **10:** 782–790.

46. Fujita, R. & H. Ueda. 2003. Protein kinase C-mediated cell death mode switch induced by high glucose. *Cell Death Differ* **10:** 1336–1347.

47. Kaplan, D.R. & F.D. Miller. 2000. Neurotrophin signal transduction in the nervous system. *Curr. Opin. Neurobiol.* **10:** 381–391.

48. Ay, I., H. Sugimori & S.P. Finklestein. 2001. Intravenous basic fibroblast growth factor (bFGF) decreases DNA fragmentation and prevents downregulation of Bcl-2 expression in the ischemic brain following middle cerebral artery occlusion in rats. *Brain Res. Mol. Brain Res.* **87:** 71–80.

49. Patapoutian, A. & L.F. Reichardt. 2001. Trk receptors: mediators of neurotrophin action. *Curr. Opin. Neurobiol.* **11:** 272–280.

50. Sofroniew, M.V., C.L. Howe & W.C. Mobley. 2001. Nerve growth factor signaling, neuroprotection, and neural repair. *Annu. Rev. Neurosci.* **24:** 1217–1281.

51. Yamashita, T., K. Sawamoto, S. Suzuki, *et al.* 2005. Blockade of interleukin-6 signaling aggravates ischemic cerebral damage in mice: possible involvement of Stat3 activation in the protection of neurons. *J. Neurochem.* **94:** 459–468.

52. Yoshida, T., I. Tomioka, T. Nagahara, *et al.* 2004. Bax-inhibiting peptide derived from mouse and rat Ku70. *Biochem. Biophys. Res. Commun.* **321:** 961–966.

53. Fujita R. & H. Ueda. 2007. Prothymosin-alpha1 prevents necrosis and apoptosis following stroke. *Cell Death Differ* **14:** 1839–1842.

54. Fujita, R., M. Ueda, K. Fujiwara & H. Ueda. 2009. Prothymosin-alpha plays a defensive role in retinal ischemia through necrosis and apoptosis inhibition. *Cell Death Differ* **16:** 349–358.

55. Stevens, S.L., P.Y. Leung, K.B. Vartanian, *et al.* 2011. Multiple preconditioning paradigms converge on interferon regulatory factor-dependent signaling to promote tolerance to ischemic brain injury. *J. Neurosci.* **31:** 8456–8463.

56. Vartanian, K.B., S.L. Stevens, B.J. Marsh, *et al.* 2011. LPS preconditioning redirects TLR signaling following stroke: TRIF-IRF3 plays a seminal role in mediating tolerance to ischemic injury. *J. Neuroinflammation* **8:** 140.

57. Leung, P.Y., S.L. Stevens, A.E. Packard, *et al.* 2012. Toll-like receptor 7 preconditioning induces robust neuroprotection against stroke by a novel type I interferon-mediated mechanism. *Stroke* **43:** 1383–1389.

58. Marsh, B., S.L. Stevens, A.E. Packard, *et al.* 2009. Systemic lipopolysaccharide protects the brain from ischemic injury by

reprogramming the response of the brain to stroke: a critical role for IRF3. *J. Neurosci.* **29:** 9839–9849.

59. Fan, H., B. Zingarelli, O.M. Peck, *et al.* 2005. Lipopolysaccharide- and gram-positive bacteria-induced cellular inflammatory responses: role of heterotrimeric Galpha(i) proteins. *Am. J. Physiol. Cell Physiol.* **289:** C293–C301.

60. Cuschieri, J., J. Billgren & R.V. Maier. 2006. Phosphatidylcholine-specific phospholipase C (PC-PLC) is required for LPS-mediated macrophage activation through CD14. *J. Leukoc. Biol.* **80:** 407–414.

61. Dauphinee, S.M., V. Voelcker, Z. Tebaykina, *et al.* 2011. Heterotrimeric Gi/Go proteins modulate endothelial TLR signaling independent of the MyD88-dependent pathway. *Am. J. Physiol. Heart Circ. Physiol.* **301:** H2246–H2253.

62. Grünberg, E., K. Eckert, H.R. Maurer, *et al.* 1997. Prothymosin alpha1 effects on IL-2-induced expression of LFA-1 on lymphocytes and their adhesion to human umbilical vein endothelial cells. *J. Interferon. Cytokine Res.* **17:** 159–165.

63. Heidecke, H., K. Eckert, K. Schulze-Forster & H.R. Maurer. 1997. Prothymosin alpha 1 effects in vitro on chemotaxis, cytotoxicity and oxidative response of neutrophils from melanoma, colorectal and breast tumor patients. *Int. J. Immunopharmacol.* **19:** 413–420.

64. Moody, T.W., J. Leyton, F. Zia, *et al.* 2000. Thymosinalpha1 is chemopreventive for lung adenoma formation in A/J mice. *Cancer Lett.* **155:** 121–127.

65. Romani, L., F. Bistoni, C. Montagnoli, *et al.* 2007. Thymosin alpha1: an endogenous regulator of inflammation, immunity, and tolerance. *Ann. N.Y. Acad. Sci.* **1112:** 326–338.

Ann. N.Y. Acad. Sci. ISSN 0077-8923

ANNALS OF THE NEW YORK ACADEMY OF SCIENCES
Issue: *Thymosins in Health and Disease*

Thymosin β4 is rapidly internalized by cells and does not induce intracellular Ca^{2+} elevation

Czeslaw S. Cierniewski,[1,2] Katarzyna Sobierajska,[1] Anna Selmi,[1] Jakub Kryczka,[1] and Radoslaw Bednarek[1]

[1]Department of Molecular and Medical Biophysics, Medical University of Lodz, Lodz, Poland. [2]Institute of Medical Biology, Polish Academy of Science Lodz, Poland

Address for correspondence: Czeslaw S. Cierniewski, Department of Molecular and Medical Biophysics, Medical University of Lodz, 92-215 Lodz, 6/8 Mazowiecka str., Poland. czeslaw.cierniewski@umed.lodz.pl

Thymosin β4 (Tβ4) is a multifunctional protein that has pleiotropic activities both intracellularly and extracellularly. The mechanisms by which it influences cellular processes such as adhesion, migration, differentiation, or apoptosis are not yet understood. Calcium is a ubiquitous signal molecule that is involved in the regulation of almost all cellular functions. Our data indicate that the release of Ca^{2+} from intracellular stores following stimulation of cells with Tβ4 does not occur. Interestingly, Tβ4 becomes rapidly internalized, supporting the concept that it may express its activities via intracellular receptors.

Keywords: thymosin β4; biological activities; calcium influx; internalization

Introduction

Thymosin β4 (Tβ4) is a typical intracellular polypeptide that, for some time now, has been considered to function solely as the major monomeric G-actin–sequestering protein in the cytoplasm. Hence, the mechanism by which it influenced cell proliferation, migration, and differentiation was generally believed to be linked with maintaining a dynamic equilibrium between G-actin and F-actin critical for the rapid reorganization of the cytoskeleton. This view was consistent with the intense staining of Tβ4 near the plasma membrane and with the function of cytoplasmic actin and actin-binding proteins known to determine multiple cellular processes.[1,2]

For the last two decades, an increasing number of studies have provided evidence to highlight the activities of Tβ4, which were not dependent upon its effect on actin polymerization.[3,4] High concentrations and ubiquitous presence of Tβ4 in the tissues and extracellular compartments suggest that Tβ4 may be an important intracellular mediator when either released from the cells or exogenously added. This 43-amino acid oligopeptide, despite the lack of a signal sequence for secretion can be detected outside of cells, in blood plasma or in wound fluid.

Extracellular concentrations of Tβ4 can be significantly increased on activation of cells by different agonists or inflammatory stimuli.[5–7] Its activities have been broadly observed as a secreted peptide in wound and oral fluids.[8,9] Furthermore, it appeared to be an essential paracrine factor in mesenchymal stem cells and embryonic endothelial progenitor cells,[10,11] although its cell surface receptors are still not known. When externally added, Tβ4-stimulated adhesion and spreading of fibroblasts,[12] differentiation of endothelial cells,[12] directional migration of endothelial cells and keratinocytes,[13] angiogenesis,[13,14] wound healing,[14,15] hair follicle growth,[16,17] apoptosis,[18] and has been described to possess anti-inflammatory properties.[16,19]

Recent studies have shown that Tβ4 is frequently overexpressed in malignant tumors and increases tumor growth, metastasis, and epithelial–mesenchymal transition (EMT).[6,18,20] It has also been reported that overexpression of Tβ4 is associated with increased invasion and distant metastasis of human colon cancer.[18] Overexpression of Tβ4 leads to (1) reduced E-cadherin level, accumulation of β-catenin in nucleus, and activation of the Tcf/LEF pathway, resulting in invasive phenotype of CRC cells;[18] (2) decreased expression of Fas and susceptibility to FasL-dependent apoptosis;[18,21]

doi: 10.1111/j.1749-6632.2012.06685.x

Ann. N.Y. Acad. Sci. 1269 (2012) 44–52 © 2012 New York Academy of Sciences.

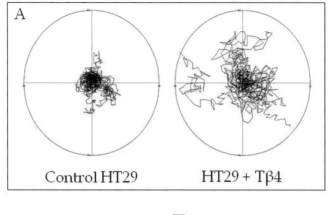

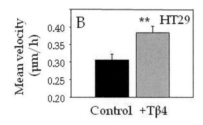
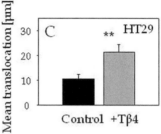

Figure 1. Increased motility of colon cancer cells treated with Tβ4. Migration of HT29 cells was tested by time-lapse microscopy. Control HT29 cells were treated with Tβ4 (160 nM) were seeded on collagen or plastic. Cell migration was followed for 24 h using a single-cell tracking software. The trajectory of 10 representative cells was analyzed. Scale bar: 180 μm. The average translocation and velocity displayed on diagrams, were determined by means of computer-assisted phase contrast videomicroscopy during a 24-h period. Data are shown as a mean of at least three determinations ± SD. **$P < 0.01$. (For additional details, see Ref. 48.)

(3) upregulation and increased secretion of PAI-1;[6] and (4) increased expression and activity of MMPs.[6] Furthermore, colon cancer HT29 cells treated with Tβ4 show enhanced random migration (Fig. 1A) as indicated by significantly higher migration speed (Fig. 1B) and average translocation (Fig. 1C) when compared to control HT29 cells (for additional details, see Ref. 48).

Tβ4 is now known to promote wound healing, tissue regeneration, and cytoprotection in the cornea, heart, skin, gingiva, and nervous system.[11,22]

It remains unclear whether Tβ4 effects are mediated by extracellular or intracellular receptors, or if Tβ4 is taken up by cells and its activity manifest after interaction with G-actin and modulation of the actin filament system.

Thymosin β4 and intracellular free Ca²⁺ concentration

Cellular events such as adhesion and migration that are stimulated by Tβ4 usually are associated with intracellular Ca²⁺ elevation.[23] The latter activates the transcription of a number of immediate early genes required for the induction of cell proliferation and motility.[24] Moreover, an increase in $[Ca^{2+}]_i$ frequently leads to nitric oxide production, which in turn promotes the proliferation of endothelial cells during wound healing.[25] Growing evidence shows that enhanced Ca²⁺ signaling contributes to carcinogenesis, increased migration and proliferation, decreased apoptosis, dedifferentiation, metastasis, and therapy resistance.[26] Remodeling of Ca²⁺ influx pathways, including elevated store-operated Ca²⁺ entry, is associated with EMT implicated in cancer metastasis. EMT involves the degradation of cell–cell and cell–extracellular matrix adhesions and the subsequent downregulation of junctional proteins such as E-cadherin.[27] Cells undergo a reorganization of the cytoskeleton and production of the type III intermediate filament vimentin,[28] leading to a change in cell shape from an epithelial to a mesenchymal or fibroblast-like morphology.[29] In breast cancer cells, EMT is associated with a remodeling of purinergic receptor–mediated Ca²⁺

signaling[30] and with increased store-operated Ca^{2+} influx.[31] Furthermore, the Ca^{2+}-related proteins, for example Orai1 and STIM1, seem to be important for store-operated calcium entry pathways in cancer cell migration and metastasis.[32]

Elevation of the cytoplasmic Ca^{2+} concentration can result from Ca^{2+} influx from the extracellular compartments or from Ca^{2+} release from intracellular stores. The main intracellular Ca^{2+} stockpile is the endoplasmic reticulum, whose Ca^{2+} content is regulated by transmembrane proteins involving the endoplasmic reticulum Ca^{2+} ATPase and the inositol 1,4,5-trisphosphate receptor. In endothelial cells, Ca^{2+} signals can be evoked by a number of mechanisms. Many agonists, such as nucleotides, acetylcholine, and growth factors, are able to trigger phospholipase C activation, inositol 1,4,5-trisphosphate production and Ca^{2+} release from the endoplasmic reticulum. In the absence of Ca^{2+} influx from the extracellular medium, a short-duration $[Ca^{2+}]_i$ increase occurs.[25] This mode of Ca^{2+} inflow, which depends on the physical coupling between the STIM protein on the Ca^{2+} store and the plasma membrane Ca^{2+} channels,[26,33–35] is the predominant pathway of Ca^{2+} entry in nonexcitable cells, including endothelial cells.[36–39] Additional routes for Ca^{2+} influx may be provided by (1) receptor-activated cation channels, activated by intracellular second messengers, such as diacylglycerol;[40] (2) mechanosensitive Ca^{2+}-permeable channels, activated by stretch, pressure, and shear stress;[40] and (3) L- and T-like voltage-dependent Ca^{2+} channels, whose expression in endothelial cells is rather limited.[41]

Recently, a novel mechanism explaining the numerous extracellular activities of Tβ4 has been proposed.[42] Tβ4 was identified to interact in endothelial cells with ecto–ATP synthase and to increase cell surface ATP levels. Consequently, the extracellular signaling pathway involving the ATP-responsive P2×4 receptor leading to enhanced HUVEC migration was proposed. P2×4 belongs to the seven mammalian P2X receptor family members that are ATP-gated ion channels,[43,44] and all these receptors, with the exception of P2×5, can facilitate entry of Ca^{2+} in response to stimulation by extracellular ATP.[45] Therefore, if this mechanism is correct, one would expect that upon treatment of endothelial cells with Tβ4, intracellular Ca^{2+} elevation should be observed. In fact, early published

observations indicated that Tβ4 could increase intracellular Ca^{2+} concentration in HL-60 leukemic cells by stimulating the release of Ca^{2+} from the intracellular pool.[45,46] The calcium signal appeared to be very weak; it started from about 40 nmol (in resting cells) and, after activation with Tβ4, reached 116 nmol (in calcium-free medium). Normally, it is more than 10-fold higher than that in cells activated with different agonists. Importantly, Tβ4 could not elicit an extracellular calcium influx.[47]

Recently, we carefully looked at the intracellular calcium influx in several cells.[48] In the course of our studies, we used two fluorescent probes, Fluo-4 NW and Fura-2AM, to monitor the Ca^{2+} signals in endothelial cells, EA.hy 926 and HT29 cells. Two potential activation pathways were considered: the release of Ca^{2+} by Tβ4-generated secondary messengers or due to the direct action of Tβ4 on the membrane $[Ca^{2+}]_i$ store. When the HUVECs were treated with trypsin, they showed an initial peak and a subsequent sustained phase in $[Ca^{2+}]_i$. The peak was caused by the Ca^{2+} release from the intracellular Ca^{2+} stores, and the sustained phase was due to the influx of extracellular Ca^{2+} across the cell membrane (Fig. 2, for additional details, see Ref. 48). A similar calcium mobilization response was observed when the HUVECs were treated with thrombin reacting with a Gq protein-coupled receptor (Fig. 2, for additional details, see Ref. 48). In contrast, HUVECs treated with 200 nM Tβ4 or its mutants (Tβ4$_{KLKKTET/7A}$, Tβ4$_{AcSDKPT/4A}$) did not show any intracellular Ca^{2+} elevation, thus indicating that the Ca^{2+} influx was not induced by these peptides (Fig. 2, for additional details, see Ref. 48). There was no visible effect even when higher concentrations of Tβ4, up to 1 μM, were used. Therefore, it seems that the Tβ4-induced intracellular Ca^{2+} elevation described by Huang *et al.*[47] could be caused by mechanosensitive Ca^{2+}-permeable channels, activated by shear stress induced during the mixing and adding of reagents to the cell culture. On the other hand, a distinct calcium response could result from different properties of cells used in both studies. In contrast to adhesive endothelial cells and colon cancer cells, HL-60 (human promyelocytic leukemia) cells are nonadherent cells that proliferate continuously in suspension culture.

There are two bioactive fragments of Tβ4, namely the N-terminal tetrapeptides AcSDKP and LKKTET, which are derived from its central

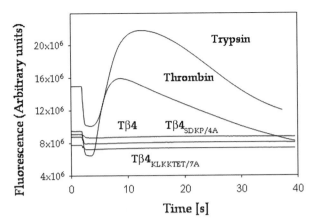

Figure 2. Tβ4 has no effect on the intracellular Ca^{2+} concentration in HUVECs. Cells grown in 96-well plates loaded with Fluo-4 NW were treated with 200 nM Tβ4 and its mutants Tβ4$_{SDKP/4A}$, Tβ4$_{KLKKTET/7A}$, or trypsin or thrombin. The kinetic data of Ca^{2+} response to trypsin, thrombin, Tβ4, and its mutants were monitored using Fluo-4 NW by the EnVision 2103 spectrofluorometer. The experiments were performed in a Ca^{2+}-containing medium. Data shown are representative of three independent experiments. (For additional details, see Ref. 48.)

actin-binding domain.[3] The present observations, in which purified mutants of Tβ4 were used, considered together with our previous studies utilizing vectors expressing the same mutants,[1] demonstrate that increased endothelial cell motility does not require the presence of these two active motifs of Tβ4. Therefore, our data indicate that the release of Ca^{2+} from intracellular stores, following stimulation of HUVECs with Tβ4, does not occur, and the data further confirm that Tβ4 may exert its effects in multiple ways: (1) directly on the actin cytoskeleton as a monomer-sequestering protein to alter actin dynamics; (2) indirectly via activation and transcription of signaling molecules, which alter the actin cytoskeleton; and (3) completely independent of actin but *via* specific intracellular receptors.

Tβ4 is involved in receptor-mediated mechanisms

Recent observations indicate that Tβ4 has activity toward different cells via receptor-mediated mechanisms.

When added externally to endothelial cells, Tβ4 induces expression and release of plasminogen activator inhibitor type 1 (PAI-1) by a mechanism involving activation of the mitogen-activated protein kinase cascade.[5,50] Tβ4, acting via an unknown receptor, stimulated JNK1 kinase activity rapidly and potently in endothelial cells, resulting in phosphorylation of Ser63 on c-Jun, a component of AP-1 transcription factor.[5] Tβ4 induced

binding of c-Fos/p-c-Jun to the highly conserved and unique AP-1–like element in the PAI-1 gene [-31], a proximal region that has been implicated in the PAI-1 transcriptional response to transforming growth factor β (TGF-β) and to PMA in several cell types. Therefore, Tβ4 appears to stimulate the PAI-1 promoter in a manner similar to most PAI-1–activating cytokines, by interacting with a common DNA-binding site, the PMA-responsive element (tPA responsive element (TRE)), and activating gene transcription in response to activators of protein kinase C (PKC), growth factors, and cytokines. Upregulation of PAI-1 expression by Tβ4 was accompanied by binding of Tβ4 to the C-terminal arm of the Ku80 subunit of ATP-dependent DNA helicase II.[1] Mutant deletion experiments revealed that the presence of the N-terminal α/β domain of Ku80 is indispensable for the effective activation of the PAI-1 gene.

Tβ4-mediated activation of ERK1/2 in endothelial cells was observed to be much lower than activation of JNK1.[5] On the other hand, ERK activation was detected in HeLa cervical cancer cells overexpressing Tβ4.[50] This may play a role in maintaining a low paclitaxel-induced cell death rate ("paclitaxel resistance"). Tβ4 inhibits paclitaxel-mediated apoptosis and induces paclitaxel resistance through the increase of basal ERK activation and reduction of p38 phosphorylation. Therefore, the basal activity of ERK and p38 kinase plays a crucial role to decide cell death or survival. Furthermore, ERK activation by Tβ4 stabilizes HIF-1α, which results in inhibition of HeLa cervical cancer cell death caused by paclitaxel.[51] Activation of ERKs was also observed after the treatment of endothelial progenitor cells (EPCs) with Tβ4, which resulted in a time- and concentration-dependent phosphorylation of Akt and endothelial nitric oxide synthase (eNOS), suggesting that Tβ4 stimulates EPC directional migration via the phosphatidylinositol 3-kinase/Akt/eNOS signal transduction pathway.[52]

Tβ4 is translocated into the nucleus by an active transport mechanism requiring an unidentified soluble cytoplasmic factor.[1,53] The amino acid sequence of Tβ4 does not contain a canonical nuclear localization signal but rather a cluster of positively charged amino acid residues ([14]KSKLKK[19]), suggestive of a functional nuclear localization signal, which partially overlaps with the proposed actin-binding site of Tβ4 ([17]LKKTET[22]). In nuclei of the human cervical cancer SiHa cell line Tβ4 was found to increase the level of zyxin,[54] a zinc-binding phosphoprotein that concentrates at focal adhesions and along the actin cytoskeleton and contains a functional nuclear export signal (NES).[55] Both zyxin and Tβ4 localized to the nucleus; moreover, Tβ4 was shown to activate translocation of zyxin into the nucleus.[54] Zyxin has been reported to play a role in actin polymerization,[55] while Tβ4 is involved in actin depolymerization; therefore Tβ4 may influence cervical cancer cell migration via zyxin by regulating actin depolymerization and polymerization events.[54]

The mechanism by which Tβ4 activates cancer cell migration and invasiveness has been examined in the colon cancer cell model.[56] Tβ4 induced EMT of SW480 colon cancer cells, which was accompanied by a loss of the epithelial marker E-cadherin, a potent inhibitor of tumor invasion, as well as a cytosolic accumulation of its associated protein, β-catenin.[56] E-cadherin downregulation seemed to be accounted by a ZEB1-mediated transcriptional repression, whereas the accumulation of β-catenin was a result of glycogen synthase kinase-3β inactivation mediated by integrin-linked kinase (ILK), which activated (by Ser473 phosphorylation) its downstream effector, Akt kinase. The extracellular domain of E-cadherin attaches the adhesion complex to the actin cytoskeleton by binding with catenins through its intracellular domain.[57] ILK upregulation by Tβ4 was triggered by an augmented stabilization of this kinase by complexing with the particularly interesting new Cys-His protein (PINCH). Moreover, overexpression of Tβ4 in SW480 colon cancer cells resulted in higher Rac1 activities and expression levels of IQ motif containing GTPase activating protein 1 (IQGAP1) and ILK.[58] In addition, IQGAP1 formed a complex with ILK, which induced the migration ability of colon cancer cells. Thus, Tβ4 increases migration of colon cancer cells via elevating the expression of IQGAP1, which forms complexes with ILK, leading finally to activation of Rac1. The upregulating effect of Tβ4 on ILK was also observed in cardiac cells.[20,59]

The anti-inflammatory events related to NF-κB suppression by Tβ4 were also evaluted in corneal cells. Tβ4 treatment decreased nuclear NF-κB protein levels, NF-κB activity, and p65 subunit phosphorylation in corneal epithelial cells after TNF-α stimulation.[60] Moreover, Tβ4 blocked nuclear translocation of the NF-κB p65 subunit in TNF-α-stimulated corneal epithelial cells. TNF-α initiates cell-signaling pathways that converge on the activation of NF-κB, thus both are known mediators of the inflammatory process. Therefore, Tβ4 inhibits the activation of NF-κB in TNF-α–stimulated cells, which results in suppression of inflammation.

Furthermore, topical Tβ4 treatment promoted corneal clarity in a BALB/c model of alkali injury.[61] The observed healing effect of Tβ4 on the cornea was associated with a decrease in polymorphonuclear cells (PMN) infiltration, downregulation of two key murine PMN chemokines—KC (murine interleukin-8–like chemokine) and macrophage inflammatory protein (MIP)-2—and a decrease in the expression of three major matrix metalloproteinases (MMPs)—the gelatinases (MMP-2, MMP-9) and leukolysin/MT6-MMP (MMP-25), a GPI-anchored MMP. The ability of Tβ4 treatment to promote corneal clarity and decrease corneal chemokine and MMP expression after alkali injury indicates that Tβ4 acts as a key regulatory wound-healing agent. Tβ4 stimulated corneal epithelial cell migration— a wound healing-related process—by induction of both transformin growth factor β1 (TGF-β1) and laminin-5 (LM-5) γ2 chain expression.[62] Tβ4 was found to increase LM-5 γ2 chain expression by the non-TGF-β1–dependent pathway, since

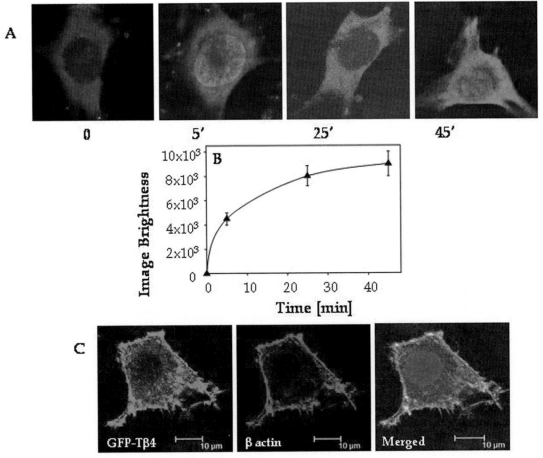

Figure 3. Tβ4 is rapidly internalized in cells. (A) The internalization of Tβ4 by EA.hy926 cells incubated with biotinylated Tβ4 for 0–45 min, followed by avidin conjugated with fluorescein. These cells are representative of a number of cells analyzed during three separate experiments. The image brightness (B) of the samples taken from sections close to the cell edges ($n = 30$) from different samples was measured using the ImageJ program. It reflects the concentration of the biotinylated Tβ4 uptaken by cells. (C) Subcellular localization of GFP-tagged Tβ4 (green) in EA.hy 926 cells transiently transfected with 5 μg of pEGFP-N1-Tβ4. Cells were fixed with paraformaldehyde; and then counterstained with TRITC-phalloidin (*red*) or Hoechst 33258 (*blue*) and signals were visualized directly by confocal microscopy. Average fluorescence derived from GFP-tagged Tβ4 or its mutants and actin filaments, when merged, revealed cellular regions with colocalization of both proteins (*yellow*). These cells are representative of a number of cells analyzed during five separate experiments. (For additional details, see Refs. 1 and 48.)

suppression of TGF-β1 further increased LM-5 γ2 chain expression.

Internalization of Tβ4 by cells

All of the above observations suggest that Tβ4 has activity on different cells *via* receptor-mediated mechanisms. Indeed, several studies focused on searching for its putative cellular receptor resulted in identifying more than a few proteins that, upon interaction with Tβ4, start to modulate important cell functions. Thus, fo-

cal adhesion proteins, such as PINCH-1 and ILK, hMLH1, Ku80, and stabilin-2 were identified to form a complex with Tβ4.[1,22,63,64] Since all these proteins are located inside cells, externally added Tβ4 must first be internalized to form a complex with the proteins and thus to modulate important cell functions. We showed that Tβ4 appears to be rapidly internalized in endothelial cells and HT29 cells.[48] Figure 3A shows the uptake of Tβ4 by EA.hy926 cells incubated with biotinylated Tβ4 for 0–45 min and detected by staining with avidin-conjugated fluorescein (for additional details, see Ref. 48). A Tβ4 influx was observed after five minutes of incubation; this is clearly seen when fluorescence intensity was quantitated using the ImageJ program (Fig. 3B, for additional details, see Ref. 48). Besides the accumulation visualized within the cytoplasm, the data obtained also showed, in many cases, punctuate nuclear staining, indicating accumulation of transported Tβ4 within the cell nucleus. Interestingly, the distribution of internalized Tβ4 within EA.hy29 cells differed from that observed when the same cells were transfected with pEGFP-N1-Tβ4, which showed much accumulation in cytoplasm close to the membranes (Fig. 3C, for additional details, see Ref. 1).

The mechanism of Tβ4 transport from the extracellular compartment to the cell cytoplasm is now under investigation. Previous reports showed that the transmembrane exchange of polypeptides can be done through several different independent mechanisms. Tβ4 fulfills all criteria of cell-penetrating peptides (CPPs), a group of short, water-soluble and partly hydrophobic and/or polybasic peptides (from 5 to 40 amino acids residues) with a net positive charge at physiological pH.[65] They have the ability to penetrate the cell membrane at low micromolar concentrations without using any chiral receptors or causing membrane damage.[65,66] However, the mechanism by which CPPs are transported is still not fully understood and is a matter of discussion.[67] There are at least two proposed cell entry mechanisms, endocytosis and direct penetration, probably via a transient pore formation. The choice of a transport mechanisms depends on many parameters, including size of peptides, cell type, temperature, and other conditions.[68,69] It was suggested that a peptide could pass through the plasma membrane via an energy-independent manner either by formation of micro-micelles at the membrane[70] or by direct translocation through the lipid bilayer.[71] CPPs with a high content of cationic residues can absorb at the cell surface to the numerous anionic moieties present in the cell membrane, such as sialic or phospholipic acids and heparin sulfate.[72] Subsequent transport by caveolae[73] and macropinocythosis[74] through a clathrin-dependent pathway,[75] and via a cholesterol-dependent, clathrin-mediated pathway,[76] has been reported. Recent observations suggest that retrograde transport can also be involved in cytosolic entry of some CPPs.[77] However,

retrograde transport in the recycling endosomes is still poorly described, and only a few reports indicate that Rab11 is probably involved in the sorting machinery.[78,79]

Acknowledgments

This work was supported by the National Science Center Project N301 4392 38.

Conflicts of interest

The authors declare no conflicts of interest.

References

1. Bednarek, R., J. Boncela, K. Smolarczyk, *et al.* 2008. Ku80 as novel receptor for thymosin β4 that mediates its intracellular activity different from G-actin sequestering. *J. Biol. Chem.* **283:** 1534–1544.
2. Pollard, P.D. & J.A. Cooper 2009. Actin, a central player in cell shape and movement. *Science* **326:** 1208–1212.
3. Goldstein, A.L., E. Hannappel & H.K. Kleinman. 2005. Thymosin β4: actin-sequestering protein moonlights to repair injured tissues. *Trends Mol. Med.* **11:** 421–429.
4. Kupatt, C.I., I. Bock-Marquette & P. Boekstegers. 2008. Embryonic endothelial progenitor cell-mediated cardioprotection requires thymosin β4. *Trends Cardiovasc. Med.* **18:** 205–210.
5. Al-Nedawi, K.N., M. Czyz, R. Bednarek, *et al.* 2004. Thymosin β4 induces the synthesis of plasminogen activator inhibitor 1 in cultured endothelial cells and increases its extracellular expression. *Blood* **103:** 1319–13242.
6. Cierniewski, C.S., I. Papiewska-Pajak, M. Malinowski, *et al.* 2010. Thymosin β4 regulates migration of colon cancer cells by a pathway involving interaction with Ku80. *Ann. N.Y. Acad. Sci.* **1194:** 60–71.
7. Huang, C.M., N. Honnavara, A.S. Barnes, *et al.* 2006. Mass spectrometric proteomics profiles of in vivo tumor secretomes: capillary ultrafiltration sampling of regressive tumor masses. *Proteomics* **6:** 6107–6116.
8. Huang, C.M., C.C. Wang, C.A Elmets, *et al.* 2006. In vivo detection of secreted proteins from wounded skin using capillary ultrafiltration probes and mass spectrometric proteomics. *Proteomics* **6:** 5805–5814.
9. Bodendorf, S., G. Born & E. Hannappel. 2007. Determination of thymosin β4 and protein in human wound fluid after abdominal surgery. *Ann. N.Y. Acad. Sci.* **1112:** 418–424.
10. Gnecchi, M., H. He, N. Noiseux, *et al.* 2006. Evidence supporting paracrine hypothesis for Akt-modified mesenchymal stem cell-mediated cardiac protection and functional improvement. *FASEB J.* **20:** 661–669.
11. Hinkel, R., C. El-Aouni, T. Olson, *et al.* 2008. Thymosin β4 is an essential paracrine factor of embryonic endothelial progenitor cell-mediated cardioprotection. *Circulation* **117:** 2232–2240.
12. Golla, R., N. Philp, D. Safer, *et al.* 1997. Co-ordinate regulation of the cytoskeleton in 3T3 cells overexpressing thymosin β4. Cell Motil. *Cytoskeleton* **38:** 187–200.

13. Grant, D.S., W. Rose, C. Yaen, *et al.* 1999. Thymosin β4 enhances endothelial cell differentiation and angiogenesis. *Angiogenesis* **3:** 125–135.

14. Malinda, K.M., G.S. Sidlu, H. Mani, *et al.* 1999. Thymosin β4 accelerates wound healing. *J. Investig. Dermatol.* **113:** 364–368.

15. Cha, H.J., M.J. Jeong & H.K. Kleinman. 2003. Role of thymosin β4 in tumor metastasis and angiogenesis. *J. Natl. Cancer Inst.* **95:** 1674–1680.

16. Sosne, G., E.A. Szliter, R. Barrett, *et al.* 2002. Thymosin β4 promotes corneal wound healing and decreases inflammation in vivo following alkali injury. *Exp. Eye Res.* **74:** 293–299.

17. Philp, D., M. Nguyen, B. Scheremeta, *et al.* 2004. Thymosin β4 increases hair growth by activation of hair follicle stem cells. *FASEB J.* **18:** 385–387.

18. Wang, W.S., P.M. Chen, H.L. Hsiao, *et al.* 2004. Overexpression of the thymosin β4 gene is associated with increased invasion of SW480 colon carcinoma cells and the distant metastasis of human colorectal carcinoma. *Oncogene* **23:** 6666–6671.

19. Qiu, P., M.K. Wheater, Y. Qiu, *et al.* 2011. Thymosin β4 inhibits TNF-a-induced NF-κB activation, IL-8 expression, and the sensitizing effects by its partners PINCH-1 and ILK. *FASEB J.* **25:** 1815–1826.

20. Huang, H.C., C.H. Hu, M.C. Tang, *et al.* 2007. Thymosin β4 triggers an epithelial-mesenchymal transition in colorectal carcinoma by upregulating integrin-linked kinase. *Oncogene* **26:** 2781–2790.

21. Hsiao, H.L. & W.S. Wang. 2006. Overexpression of thymosin β4 renders SW480 colon carcinoma cells more resistant to apoptosis triggered by FasL and two topoisomerase II inhibitors via downregulating Fas and upregulating surviving expression, respectively. *Carcinogenesis* **27:** 936–944.

22. Bock-Marquette, I., A. Saxena, M.D. White, *et al.* 2004. Thymosin β4 activates integrin-linked kinase and promotes cardiac cell migration, survival and cardiac repair. *Nature* **432:** 466–447.

23. Clapham, D.E. 2007. Calcium signaling. *Cell* **131:** 1047–1058.

24. Tran, P.O., L.E. Hinman, G.M. Unger, *et al.* 1999. A wound induced [Ca2+]i increase and its transcriptional activation of immediate early genes is important in the regulation of motility. *Exp. Cell Res.* **246:** 319–326.

25. Erdogan, E., C.A. Schaefer, M. Schaefer, *et al.* 2005. Margatoxin inhibits VEGF-induced hyperpolarization, proliferation and nitric oxide production of human endothelial cells. *J. Vasc. Res.* **42:** 368–376.

26. Bergner, A. & R.M. Huber. 2008. Regulation of the endoplasmic reticulum Ca2+-store in cancer. *Anti-Cancer Agents Med. Chem.* **8:** 705–709.

27. Polyak, K. & R.A. Weinberg. 2009. Transitions between epithelial and mesenchymal states: acquisition of malignant and stem cell traits. *Nat. Rev. Cancer* **9:** 265–273.

28. Gilles, C., M. Polette, J.M. Zahm, *et al.* 1999. Vimentin contributes to human mammary epithelial cell migration. *J. Cell Sci.* **112:** 4615–4625.

29. Jo, M., R.D. Lester, V. Montel, *et al.* 2009. Reversibility of epithelial-mesenchymal transition (EMT) induced in breast cancer cells by activation of urokinase receptor-dependent cell signaling. *J. Biol. Chem.* **284:** 22825–22833.

30. Davis, F.M., P.A. Kenny, E.T.L. Soo, *et al.* 2011. Remodeling of purinergic receptor-mediated Ca2+ signaling as a consequence of EGF-induced epithelial-mesenchymal transition in breast cancer cells. *PLoS One* **6:** e23464.

31. Hu, J.J., K.H. Qin, Y. Zhang, *et al.* 2011. Downregulation of transcription factor Oct4 induces an epithelial-to-mesenchymal transition via enhancement of Ca2+ influx in breast cancer cells. *Biochem. Biophys. Res. Commun.* **411:** 786–791.

32. Yang, S., J.J. Zhang & X.Y. Huang. 2009. Orai1 and STIM1 are critical for breast tumor cell migration and metastasis. *Cancer Cell* **15:** 124–134.

33. Adams, D.J., J. Barakeh & L.C. Van Breemen. 1989. Ion channels and regulation of intracellular calcium in vascular endothelial cells. *FASEB J.* **3:** 2389–2400.

34. Ambudkar, I.S., H.L. Ong, X. Liu, *et al.* 2007. TRPC1: the link between functionally distinct store-operated calcium channels. *Cell Calcium* **42:** 213–223.

35. Cahalan, M.D., S.L. Zhang, A.V. Yeromin, *et al.* 2007. Molecular basis of the CRAC channel. *Cell Calcium* **42:** 133–144.

36. Nilius, B. & G. Droogmans. 2001. Ion channels and their functional role in vascular endothelium. *Physiol. Rev.* **81:** 1415–1459.

37. Bishara, N.B., T.V. Murphy & M.A. Hill. 2002. Capacitative Ca2+ entry in vascular endothelial cells is mediated via pathways sensitive to 2 aminoethoxydiphenyl borate and xestospongin C. *Br. J. Pharmacol.* **135:** 119–128.

38. Cioffi, D.L., S. Wu & T. Stevens. 2002. On the endothelial cell I (SOC). *Cell Calcium* **33:** 323–336.

39. Dedkova, E.N. & L.A. Blatter. 2003. Nitric oxide inhibits capacitative Ca2+ entry and enhances endoplasmic reticulum Ca2+ uptake in bovine vascular endothelial cells. *J. Physiol.* **539:** 77–91.

40. Nilius, B., G. Droogmans & R. Wondergem. 2003. Transient receptor potential channels in endothelium: solving the calcium entry puzzle? *Endothelium* **10:** 5–15.

41. Bossu, J.L., A. Feltz, J.L. Rodeau, *et al.* 1989. Voltage-dependent transient calcium currents in freshly dissociated capillary endothelial cells. *FEBS Lett.* **255:** 377–380.

42. Freeman, K.W., B.R. Bowman & B.R. Zetter. 2011. Regenerative protein thymosin b4 is a novel regulator of purinergic signaling. *FASEB J.* **25:** 907–915.

43. Vial, C., J.A. Roberts & R.J. Evans. 2004. Molecular properties of ATP-gated P2X receptor ion channels. *Trends Pharmacol. Sci.* **25:** 487–493.

44. North, R.A. 2002. Molecular physiology of P2X receptors. *Physiol. Rev.* **82:** 1013–1067.

45. Di Virgilio, F., P. Chiozzi, D. Ferrari, *et al.* 2001. Nucleotide receptors: an emerging family of regulatory molecules in blood cells. *Blood* **97:** 587–560.

46. Sluyter, R., J.A. Barden & J.S. Wiley. 2001. Detection of P2X purinergic receptors on human B lymphocytes. *Cell Tissue Res.* **304:** 231–236.

47. Huang, W.Q., B.H. Wang & Q.R. Wang. 2006. Thymosin β4 and AcSDKP inhibit the proliferation of HL-60 cells and induce their differentiation and apoptosis. *Cell Biol. Int* **30:** 514–520.

48. Selmi, A., M. Malinowski, W. Brutkowski, *et al.* 2012. Thymosin β4 promotes the migration of endothelial cells without intracellular Ca2 +elevation. *Exp. Cell Res.* **18:** 1659–1666.

49. Boncela, J., K. Smolarczyk & E. Wyroba. 2006. Binding of PAI-1 to endothelial cells stimulated by thymosin β4 and modulation of their fibrinolytic potential. *J. Biol. Chem.* **281:** 1066–1072.

50. Oh, S.Y., J.H. Song, J.E. Gil, *et al.* 2006. ERK activation by thymosin β4 (Tβ4) overexpression induces paclitaxel-resistance. *Exp. Cell Res.* **312:** 1651–1657.

51. Oh, J.M., I.J. Ryoo, Y. Yang, *et al.* 2008. Hypoxia-inducible transcription factor (HIF)-1α stabilization by actin-sequestering protein, thymosin β4 (Tβ4) in Hela cervical tumor cells. *Cancer Lett.* **264:** 29–35.

52. Qiu, F.Y., X.X. Song, H. Zheng, *et al.* 2009. Thymosin β4 induces endothelial progenitor cell migration via PI3K/Akt/eNOS signal transduction pathway. *J. Cardiovasc. Pharmacol.* **53:** 209–214.

53. Huff, T., O. Rosorius, A.M. Otto, *et al.* 2004. Nuclear localization of the G-actin sequestering peptide thymosin β4. *J. Cell Sci.* **117:** 5333–5343.

54. Moon, H.S., S. Even-Ram, H.K. Kleinman, *et al.* 2006. Zyxin is upregulated in the nucleus by thymosin β4 in SiHa cells. *Exp. Cell Res.* **312:** 3425–343

55. Beckerle, M.C. 1997. Zyxin: zinc fingers at sites of cell adhesion. *BioEssays* **19:** 949–9571.

56. Huang, H.C., C.H. Hu, M.C. Tang, *et al.* 2007. Thymosin β4 triggers an epithelial–mesenchymal transition in colorectal carcinoma by upregulating integrin-linked kinase. *Oncogene* **26:** 2781–2790.

57. Takeichi, M. 1995. Morphogenetic roles of classic cadherins. *Curr. Opin. Cell Biol.* **7:** 619–627.

58. Tang, M.C., L.C. Chan, Y.C. Yeh, *et al.* 2011. Thymosin β4 induces colon cancer cell migration and clinical metastasis via enhancing ILK/IQGAP1/Rac1 signal transduction pathway. *Cancer Lett.* **308:** 162–171.

59. Sopko, N., Y. Qin, A. Finan, *et al.* 2011. Significance of thymosin β4 and implication of PINCH-1-ILK-α-Parvin (PIP) complex in human dilated cardiomyopathy. *PLoS One* **6:** e20184.

60. Sosne, G., P. Qiu, P.L. Christopherson, *et al.* 2007. Thymosin β4 suppression of corneal NFκB: a potential anti-inflammatory pathway. *Exp. Eye Res.* **84:** 663–669.

61. Sosne, G., P.L. Christopherson, R.P. Barrett, *et al.* 2005. Thymosin β4 modulates corneal matrix metalloproteinase levels and polymorphonuclear cell infiltration after alkali injury. *Invest. Ophthalmol. Vis. Sci.* **46:** 2388–2395.

62. Sosne, G., L. Xu, L. Prach, *et al.* 2004. Thymosin β4 stimulates laminin-5 production independent of TGF-beta. *Exp. Cell Res.* **293:** 175–183.

63. Brieger, A., G. Plotz, S. Zeuzem, *et al.* 2007. Thymosin β4 expression and nuclear transport are regulated by hMLH1. *Biochem. Biophys. Res. Commun.* **364:** 731–736.

64. Lee, S.J., I.S. So, S.Y. Park, *et al.* 2008. Thymosin β4 is involved in stabilin-2-mediated apoptotic cell engulfment. *FEBS Lett.* **582:** 2161–2166.

65. Jarver, P. & U. Langel. 2006. Cell-penetrating peptides—a brief introduction. *Biochim. Biophys. Acta* **1758:** 260–263.

66. El-Andaloussi, S., T. Holm & U. Langel. 2005. Cell-penetrating peptides: mechanisms and applications. *Curr. Pharm. Design* **11:** 3597–3611.

67. Heitz, F., M.C. Morris & G. Divita. 2009. Twenty years of cell-penetrating peptides: from molecular mechanisms to therapeutics. *Br. J. Pharmacol.* **157:** 195–206.

68. Tunnemann, G., R.M. Martin & S. Haupt. 2006. Cargo-dependent mode of uptake and bioavailability of TAT-containing proteins and peptides in living cells. *FASEB J.* **20:** 1775–1784.

69. Morris, M.C., J. Depollier & J. Mery. 2001. Peptide carrier for the delivery of biologically active proteins into mammalian cells. *Nat. Biotech.* **19:** 1173–1176.

70. Derossi, D., S. Calvet & A. Trembleau. 1996. Cell internalization of the third helix of the Antennapedia homeodomain is receptor-independent. *J. Biol. Chem.* **271:** 18188–18193.

71. Thoren, P.E., D. Persson & P. Lincoln. 2005. Membrane destabilizing properties of cell-penetrating peptides. *Biophys. Chem.* **114:** 169–179.

72. Vives, E. 2003. Cellular uptake of the Tat peptide: an endocytosis mechanism following ionic interactions. *J. Mol. Recogn.* **16:** 265–271.

73. Fittipaldi, A., A. Ferrari, M. Zoppe, *et al.* 2003. Cell membrane lipid rafts mediate caveolar endocytosis of HIV-1 Tat fusion proteins. *J. Biol. Chem.* **278:** 34141–34149.

74. Nakase, I., M. Niwa, T. Takeuchi, *et al.* 2004. Cellular uptake of arginine-rich peptides: roles for macropinocytosis and actin rearrangement. *Mol. Ther.* **10:** 1011–1022.

75. Saalik, P., A. Elmquist, M. Hansen, *et al.* 2004. Protein cargo delivery properties of cell-penetrating peptides. A comparative study. *Bioconjug. Chem.* **15:** 1246–1253.

76. Foerg, C., U. Ziegler, J. Fernandez-Carneado, *et al.* 2005. Decoding the entry of two novel cell-penetrating peptides in HeLa cells: lipid raft-mediated endocytosis and endosomal escape. *Biochemistry* **44:** 72–81.

77. Fisher, R., K. Kohler, M. Fotin-Mleczek, *et al.* 2006. A stepwise dissection of the intracellular fate of cationic cell-penetrating peptides. *J. Biol. Chem.* **279:** 12625–12635.

78. Chaudhry, A., S.R. Das, S. Jameel, *et al.* 2008. HIV-1 Nef induces a Rab11-dependent routing of endocytosed immune costimulatory proteins CD80 and CD86 to the Golgi. *Traffic* **9:** 1925–1935.

79. Wilcke, M., L. Johannes, T. Galli, *et al.* 2000. Rab11 regulates the compartmentalization of early endosomes required for efficient transport from early endosomes to the trans-Golgi network. *J. Cell. Biol.* **151:** 1207–1220.

Ann. N.Y. Acad. Sci. ISSN 0077-8923

Thymosin β4 expression reveals intriguing similarities between fetal and cancer cells

Gavino Faa,[1] Sonia Nemolato,[1] Tiziana Cabras,[2] Daniela Fanni,[1] Clara Gerosa,[1] Mattia Fanari,[1] Annalisa Locci,[1] Vassilios Fanos,[3] Irene Messana,[2] and Massimo Castagnola[4]

[1]Department of Pathology, University Hospital San Giovanni di Dio, [2]Department of Life and Environmental Sciences, [3]NICU Center, University of Cagliari, Cagliari, Italy. [4]Institute of Biochemistry and Clinical Biochemistry, Faculty of Medicine, Catholic University, Rome, Italy

Address for correspondence: Daniela Fanni, Department of Pathology, University Hospital San Giovanni di Dio, University of Cagliari, Via Ospedale 54, 09124 Cagliari, Italy. fandan73@yahoo.it

Thymosin β4 (Tβ4) is highly expressed in saliva of human newborns but not in adults. Here preliminary immuno-histochemical analyses on different human tissues are reported. Immunoreactivity for Tβ4 in human salivary glands show high quantities of Tβ4 before birth, followed by downregulation of expression in adulthood. In contrast, Tβ4 is detected in tumors of salivary glands, suggesting that tumor cells might utilize fetal programs, including Tβ4 synthesis. Immunohistochemical analyses in the gastrointestinal tract showed strong reactivity for Tβ4 in enterocytes during development, but weak immunostaining in mature enterocytes. In colorectal cancer, the association of a high expression of Tβ4 with epithelial–mesenchymal transition was observed. On the basis of these data, the process of epithelial–mesenchymal transition could represent the unifying process that explains the role of Tβ4 during fetal development and in cancer progression.

Keywords: thymosin beta 4; immunohistochemistry; development; cancer; embryogenesis

Introduction

Thymosin β4 (Tβ4) is a naturally occurring peptide, first isolated in 1966 by Goldstein *et al.* from the calf thymus among other lymphocytopoietic factors.[1] The complete amino acid sequence of Tβ4 was described in 1981; it contains 43 amino acids, with a high proportion of lysyl and glutamyl residues.[2] The human Tβ4 gene (hTβ4) is located on chromosome X and comprises three exons and two introns.[3] The primary translation product is modified by removal of the N-terminal methionine and acetylation.[4,5]

Recent studies highlighted extracellular roles for Tβ4 (for a recent review, see Ref. 6). However, because the peptide does not contain N-terminal leader sequences, the mechanism of release is completely unknown.[7]

Tβ4 is considered the most abundant among β-thymosin peptides in mammalian tissues: its activity has been mainly related to the regulation of actin polymerization in living cells.[8,6] Tβ4 is also thought to be involved in many critical biological activities,[9] including angiogenesis,[10] wound healing,[11] inflammatory response,[12] and cell migration.[13] Most mammalian cells express two β-thymosin variants simultaneously; Tβ4, associated with Tβ10, is the most abundant β-thymosin in human cells.[14–16]

Our interest in Tβ4 takes its origin from the finding that this peptide is highly expressed in saliva of human newborns, but not in saliva of adults.[17] It was necessary to first determine in which tissue, and in which cells, Tβ4 is synthesized and secreted into the oral cavity. To answer these questions, we analyzed tissue samples from human salivary glands of neonates with different gestational ages and performed immunohistochemical analyses on these tissues looking for the presence of the peptide. Tβ4 immunostaining was identified in acinar cells of the parotid, submandibular, and sublingual glands, as well as in minor salivary glands, clearly indicating these cells as the source of Tβ4 in

doi: 10.1111/j.1749-6632.2012.06679.x

the saliva of human newborns.[18] These findings motivated us to study Tβ4 expression in other organs and tissues of human fetuses and adults to provide more data on the expression of the peptide in humans. We also analyzed Tβ4 immunoreactivity in various human tumors of different origin, including tumors of salivary glands and colorectal cancer.

In the present paper, our research on Tβ4 expression in human tissues in health and in disease will be summarized, particularly emphasizing the role of Tβ4 in the process of epithelial to mesenchymal transition (EMT),[19] both during human development and in tumor progression.

Tβ4 in human salivary glands

The aim of our work on Tβ4 in human salivary glands has been to (1) verify whether the peptide is secreted by acinar cells of salivary glands; (2) identify whether the pattern of Tβ4 secretion and expression change at different gestational ages during the intrauterine life; and (3) control whether the Tβ4 expression pattern could change in neonates and adult subjects. To this end, we analyzed parotid, submandibular, sublingual, and minor salivary gland tissue samples obtained from human fetuses of different gestational ages. Immunohistochemical studies clearly demonstrated the presence of two main protein reactivity patterns: a granular pattern, observed in the cytoplasm of acinar cells, inside the ductal lumen and in the connective tissues surrounding the epithelial structures; a diffuse pattern, characterized by the homogeneous staining of the entire cytoplasm, mainly detected in ductal cells.[18] We hypothesized that the granular immunoreactivity could be related to Tβ4 secretion in two ways: at the apical pole of acinar cells into saliva, in which the peptide is present in high quantities,[12,20] and at the basolateral pole into the connective tissues, in which the peptide could have autocrine or paracrine functions. The homogeneous cytoplasmic pattern, mainly found in the ductal cells of adult salivary glands, was interpreted as characteristic of the binding of Tβ4 to G-actin monomers.[18]

When we analyzed immunoreactivity for Tβ4 in tumors originating from salivary glands, we detected the peptide in the vast majority of neoplasias studied. In particular, a strong expression for the peptide was detected in mixed tumors—Tβ4 being found in the cytoplasm of myoepithelial tumor cells (Fig. 1A) and in Warthin tumor cells (Fig. 1B).

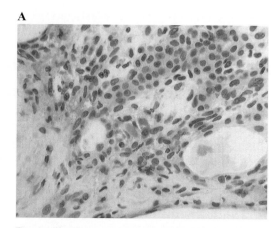

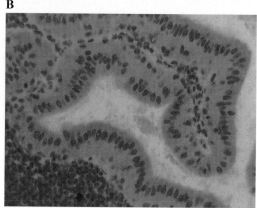

Figure 1. Tβ4 immunoreactivity in salivary gland tumors. (A) Mixed tumor of parotid gland: a strong granular and diffuse cytoplasmic immunoreactivity for Tβ4 is detected in myoepithelial tumor cells (OM ×400). (B) Warthin tumor: a strong diffuse Tβ4 positivity is observed in the cytoplasm of epithelial cells. Scattered perinuclear coarse granules are detected in the cytoplasm of the same tumoral cells (OM ×400).

Our data suggest that Tβ4 expression in human salivary glands may be summarized by (1) strong reactivity in fetal glands; (2) marked decrease in expression in adult glands; and (3) re-expression in tumor progression (Fig. 2). Moreover, in some salivary gland tumors, a great number of intra- and peritumoral mast cells were observed, all characterized by a strong immunostaining for the peptide.[21] This finding could indicate a role for Tβ4 not restricted to the physiological development of salivary glands, but also in cancer development and progression, likely due to the use of fetal programs by salivary gland cancer cells. In line with this hypothesis, the observation of Tβ4-rich tumor-infiltrating mast cells in salivary gland

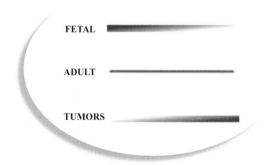

Figure 2. Changes in Tβ4 reactivity in human salivary glands in different prenatal and postnatal ages and in neoplasia.

tumors underscores the hypothesis that this peptide could serve as a local paracrine mediator, with a relevant role in cellular cross-talking within the tumor microenvironment.[22]

Concerning the function of Tβ4 in neoplastic cells, this peptide has been shown to exert anti-inflammatory and cytoprotective functions by suppressing secretion of the proinflammatory cytokine IL8 and by protecting cells against TNF-α–induced apoptosis[23] and by inhibiting neutrophil infiltration and decreasing the expression of proinflammatory cytokines.[24] Because of its multifunctional roles in protecting cells against apoptosis[17,24] and in stimulating neoangiogenesis,[25] Tβ4 released by tumor cells and/or by mast cells in the tumor microenvironment could significantly contribute to cancer cell survival and diffusion. As a consequence, Tβ4 might represent a new molecular target to be considered for future antitumor strategies[26] in different human tumors. The finding that strong expression of Tβ4 in the cytoplasm of tumor-infiltrating mast cells[21] extends our knowledge regarding the immunophenotypic profile of mast cells and contributes to our understanding of immune cell/cancer cell crosstalk. The possible relationship between the degree of Tβ4-immunreactive mast cell infiltration and tumor behavior warrants further consideration in future investigations.

Tβ4 in the human gastrointestinal tract

Tβ4 has been suggested as the ideal actin monomer-sequestering protein.[27] Its function was first restricted to regulate actin polymerization of non-muscle cells, with multiple effects on cell surface remodeling and motility.[28] Further data suggesting

a role of Tβ4 in modulating stem cell migration,[25] activation,[29] and inhibition,[30] as well as in regulating integrin signaling,[31] has prompted some authors to speak of the "β-thymosin enigma."[32]

The theory on the putative role of Tβ4 in the physiological development of embryos, as well as in vascularization and tissue recovery in acute and chronic ischemia, was reinforced by the discovery that Tβ4 is one of the most abundant factors secreted by embryonic endothelial progenitor cells.[33]

Data suggesting a role for Tβ4 in the recruitment of stem cells in different organs, and in particular during embryonic and fetal development, prompted us to investigate the expression of the peptide in human fetuses and embryos of different gestational ages, assessing a potential role of Tβ4 in the development of the different components of the gastrointestinal tract.[34] In that study, samples from gut, liver, and pancreas were analyzed for Tβ4 expression. Tβ4 was highly expressed in the epithelial cells during the early phases of the development, both in gut and pancreas, confirming previous studies on a possible relevant role of Tβ4 in the development of the gastrointestinal tract. For the first time, a marked heterogeneity of Tβ4 expression within the gastrointestinal tract was found, ranging from a diffuse immunoreactivity for the peptide in pancreas and gastrointestinal cells to the absence of the protein in the vast majority of fetal and newborn livers examined. Moreover, we detected marked differences in Tβ4 expression among different cell types within the single organs. The most striking differences were found in the fetal pancreas: Tβ4 was strongly expressed in the endocrine cells of the Langerhans islets in the absence of any significant reactivity in exocrine acinar and ductal cell. Interindividual differences were also reported regarding the intensity of the immunoreactivity for Tβ4 and its subcellular localization, primarily related to the different gestational age of the subjects studied. The strong positivity of Tβ4 in multiple cell types of the developing gastrointestinal tract in humans suggests a relevant role for the peptide in human physiological development.

When the Tβ4 expression pattern was analyzed in the adult gastrointestinal tract, we observed a marked decrease in immunoreactivity for the peptide. In particular, enterocytes of the ileum and colon did not show any significant reactivity for Tβ4, which was not detected even in intestinal

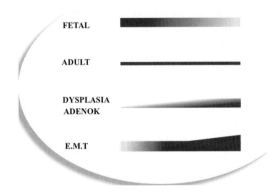

FETAL

ADULT

DYSPLASIA
ADENOK

E.M.T

Figure 3. Changes in Tβ4 reactivity in the human gastrointestinal tract in different prenatal and postnatal ages and in neoplasia.

glands (Fig. 3). The pattern of immunostaining for Tβ4 in the adult pancreas seemed similar to that described in fetal pancreas. On the contrary, significant changes were detected in the adult human liver: Tβ4 was highly expressed in the vast majority of adult hepatocytes, with a preferential localization in the hepatocytes bordering the terminal veins (zone 3 of the acinus).[35]

Tβ4 in colon cancer is highly expressed in tumor cells undergoing EMT

In colorectal cancer, Tβ4 immunoreactivity was detected in tumor cells of the vast majority of colon carcinomas studied, showing a patchy distribution, with well-differentiated areas significantly more reactive than the less differentiated tumor zones.[36] We also noted a zonal pattern in the majority of tumors, characterized by a progressive increase in immunostaining for Tβ4 from the superficial toward the deepest tumor regions. The strongest expression of Tβ4 was preferentially detected at the invasive margins of colorectal cancer, specifically invading tumor cells with features in EMT. The increase in reactivity for Tβ4 was paralleled by a progressive decrease in E-cadherin expression in cancer cells undergoing EMT. The significance of Tβ4 reactivity in colon cancer cells, however, remains unknown. The preferential expression of the peptide and the increase in intensity of the immunostaining at the invasion front of colon cancer lead to a possible link between the peptide and the process of EMT. Tβ4 may therefore have a major role in colorectal cancer

invasion and metastasis and may serve as a possible future target for anticancer chemotherapy.[26]

Taken together, our data on Tβ4 expression in the gastrointestinal tract parallel previous data obtained in human salivary glands. The changes in Tβ4 expression, from the intrauterine life to adulthood and eventually during tumor progression may be summarized as strong reactivity during fetal life, progressive silencing of Tβ4 expression in the adult gut, and reexpression of the peptide in colon cancer tumor cells (Fig. 4). These observations confirm the hypothesis that tumor cells might use, for their progression and survival, cell programs typically employed by fetal stem cells, and the synthesis of Tβ4 could be included, on the basis of our data, among the fetal programs useful for tumor progression.

The observation of the presence of Tβ4 mainly in cells undergoing EMT at the invasion front of colorectal cancer deserves consideration. EMT is a process characterized by the loss of original epithelial features in embryonic and in tumor cells, accomplished by the gain of a mesenchymal phenotype, producing nonpolarized isolated cells embedded in the extracellular matrix.[37] At the molecular level, EMT requires multiple events, such as disruption of intercellular junctions, loss of cell polarity, microtubule disruption, and basement membrane breakdown.[19] Recent studies have proposed a new model for EMT, indicating a new sequence in the distinct cellular steps that eventually lead to EMT.[38] In this model, microtubules, already known to have a central role in cell adhesion to basal membranes, cell polarity, and migration,[39] lose their stability. Microtubule disruption causes basement membrane disassembly, disruption of the epithelial cell–basal membrane interaction and, eventually, breakdown of the basal membrane.[38] According to this model, disruption of cell–cell junctions could take place only in a second iteration, followed by the proper execution of EMT and by invasion of the surrounding mesenchyma.

EMT has been originally described by embryologists as a key process in many developmental processes,[40] including the formation of the neural crest and of muscle fibers in the myotome.[41] Recently, EMT has emerged as a key step in cancer progression, allowing tumor cells to acquire invasive behavior and disseminate, originating with distant metastasis.[42,43] EMT is now believed to be a major mechanism by which cancer cells become migratory

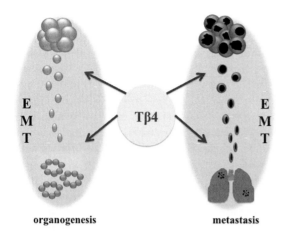

organogenesis

metastasis

Figure 4. Schematic representation of the role played by Tβ4 in fetal development and in tumor progression.

and invasive,[44] allowing tumor cells to translocate from the initial neoplastic core into neighboring host tissues, penetrate vessel endothelium, and enter circulation to form distant metastases.[45]

At the histological level, the typical features of EMT may be observed at the infiltrative margins of carcinomas as individual malignant cells with different degrees of atypia, often acquiring a spindle shape and detaching from the tumor mass, staying independently within the interstitial matrix of the peritumoral stroma.[45] At the immunohistochemical level, EMT is characterized by a dramatic decrease in intercellular expression of E-cadherin,[46] which has been postulated to result from disruption of the adherens junctions.[47]

In colorectal cancer, EMT is evidenced by a change in tumor tissue architecture at the deep invasive tumor margins. This particular modification of cancer architecture has been referred to as "budding margins"—infiltrative margins with solid cell nests formed by two to three cancer cells, which display their acquisition of motility by infiltrating the peritumoral connective tissue. Tβ4, generally associated with the regulation of actin polymerization by binding and sequestering monomeric G-actin[8] could trigger EMT in colorectal carcinoma by upregulating integrin-linked kinase (ILK).[48] Overexpression of Tβ4 has been shown to upregulate ILK,[13] causing the suppression of E-cadherin expression, and resulting in the disruption of adherens junctions and in the induction of EMT.[46]

Our data in colon carcinogenesis in humans clearly confirm previous experimental and *in vitro*

studies, indicating Tβ4 as one of the main actors in this process, strictly related to the ability of tumor cells to give rise to distant metastases. According with this hypothesis, Tβ4 should be considered as a possible target of future anticancer therapies.

Tβ4 in the human liver

A study from our group carried out on human newborn livers did not reveal any significant reactivity for Tβ4 in the vast majority of liver cells at birth.[34] On the basis of these data, and given the very high concentration of Tβ4 reported in human platelets,[49] we used immunohistochemistry to investigate the Tβ4 expression pattern in the adult normal liver, to shed light on the role played by liver cells in the synthesis of this peptide. Tβ4 reactivity was detected in all liver biopsies of adult subjects studied, with no relationship between the degree of Tβ4 immunostaining and age or sex of the subjects. Immunoreactivity for the peptide was mainly localized in the cytoplasm of hepatocytes as large granules different in shape and size. At low power, immunoreactivity for Tβ4 was not homogeneously diffused to the entire parenchyma, but it clearly showed zonation—periterminal hepatocytes (zone 3 of the acinus) showed the highest levels of Tβ4 storage, while a minor degree of reactivity for the peptide was detected in periportal hepatocytes (zone 1). A progressive increase in the degree of immunostaining from the periportal (zone 1) to the periterminal (zone 3) areas was often observed, intermediate levels of Tβ4 immunostaining being observed in the hepatocytes of acinar zone 2. The degree of reactivity changed from one case to the next. In some cases, only the periterminal hepatocytes were reactive for the peptide, whereas in other cases all the acinar zones were immunostained. Even though the vast majority of stored peptide was constantly found in the hepatocytes, Tβ4 staining was also rarely detected in scattered Kupffer cells. Ito cells did not show any reactivity for Tβ4.[35]

Whereas data on Tβ4 expression in newborn liver[34] were in favor of a minor role of human liver in Tβ4 metabolism, on the contrary, the high amounts of Tβ4 found in the adult "normal" liver indicate the hepatocyte as the most important producer of Tβ4 in the human body. The two conditions, the newborn and the adult, seem to represent the extremes of a spectrum: complete absence of the peptide in liver cells at birth, and a strong diffuse accumulation in the hepatocytes in adults.

It is unclear why this difference in the amount of Tβ4 exists between adults and newborns, but it confirms previous reports on the ability of Tβ4, as well of other β-thymosins, to change expression over the life span.[18] Our work is consistent with there being marked variability in the expression and function of Tβ4 in different organs. For example, the liver is characterized by the absence of Tβ4 at birth, and its significant synthesis and secretion in the adult.

The intracellular concentration of Tβ4 is particularly high in cells rapidly responding to external signals by profound shape changes, such as platelets,[50,51] or by increased motility, such as polymorphonuclear cells,[52] mast cells,[21] and macrophages.[53,54] The observation of large amounts of Tβ4 in hepatocytes—that is, in cells that do not rapidly change their shape or their motility—clearly support the hypothesis that Tβ4 is not a simple actin monomer–buffering protein.[28] The significant Tβ4 expression found in the adult human liver identifies the hepatocyte as a central cell in the body in which synthesis and metabolism of Tβ4 occur. Serum concentration of free Tβ4 was first determined by Naylor *et al.*[55] by a radioimmunoassay to be between 0.45–1.1 mg/L, and later by HPLC to be between 0.2 and 2 mg/L,[56] while the concentration in plasma was 0.03–0.4 mg/L. In a review by Mannherz and Hannappel,[49] the molar concentration in serum/plasma was calculated to be 10–200 nM, which is very high compared to the concentration of insulin (up to 140 pM) or glucagon (60 pM),[43] but comparable to adenosine (30–300).[57] Such a high serum level of Tβ4 in the steady state cannot be explained by its release from platelets during blood coagulation[14] or from leukocytes and monocytes during inflammation.[58] On the contrary, our work in progress suggests that the high serum content of Tβ4 could originate from secretion into the sinusoidal lumen by hepatocytes.

Conclusions

The study of Tβ4 expression in tissues from human subjects in different phases of life—ranging from the intrauterine, to the perinatal period, to adulthood, and from physiological situations to inflammatory and tumor pathology—suggests a fascinating picture characterized by marked changes in the expression of this peptide from one cell type to the next, and from one organ to the next. Interestingly, in examining a new organ or tissue, we were not able to predict these changes. In some organs, such as the liver, the absence of immunoreactivity in the prenatal and perinatal periods contrasted with the high amounts of Tβ4 observed in liver cells of adults. At the other end of the spectrum, we could find salivary glands with marked expression of Tβ4 in fetuses and newborns, but low immunoreactivity in adults. In other organs, including the pancreas, which could be placed in the middle of the spectrum, immunostaining for Tβ4 did not change with age and was constantly restricted to endocrine pancreatic cells. Marked differences in Tβ4 expression among different organs were also observed by our group in an immunohistochemical study in the genitourinary tract.[59]

The complexity of these changes made it difficult to understand the rules of the events we observed under the microscope. By using histopathological diagnosis in oncology and applying Tβ4 immunocytochemistry in tumor cells, we found that the association between Tβ4 with EMT was the most relevant. EMT could represent, in our opinion, the unifying process to explain the relevant role of Tβ4 during both fetal development and in the development of human tumors. In recent years, many authors have underlined the oncogenic potential of many transcription factors that are first expressed in a orderly manner during fetal development and that often disappear in mature tissues but may be reexpressed during neoplastic transformation.[60,61] These studies, consistent with our data, showing high Tβ4 expression in many human tissues during development, followed by downregulation in mature tissues and then reexpression in high quantities in tumor cells undergoing EMT, force us to consider the role of Tβ4 expression in human cell types. For example, Tβ4 could be seen as one of the multiple factors that propel both fetal stem/progenitor cells and tumor cells moving between different stages from epithelial to mesenchymal stages, and vice versa, allowing migration and differentiation in fetal life and metastasis in tumor progression. The observation in colon cancer cells of a strong reactivity for Tβ4 in tumor cells undergoing EMT, associated with the absence of Tβ4 immunostaining in tumor cells undergoing apoptosis, suggests a possible antiapoptotic role for Tβ4 in cancer and that Tβ4-targeted therapy might induce effective killing of colon cancer cells.

Conflicts of interest

The authors declare no conflicts of interest.

References

1. Goldstein, A.L., F.D. Slater & A. White. 1966. Preparation, assay, and partial purification of a thymic lymphocytopoietic factor (thymosin). *Proc. Natl. Acad. Sci. U. S. A.* **56:** 1010–1017.

2. Low, T.L., S.K. Hu & A.L. Goldstein. 1981. Complete amino acid sequence of bovine thymosin beta 4: a thymic hormone that induces terminal deoxynucleotidyl transferase activity in thymocyte populations. *Proc. Natl. Acad. Sci. U. S. A.* **78:** 1162–1166.

3. Yang, S.P., H.J. Lee & Y. Su. 2005. Molecular cloning and structural characterization of the functional human thymosin beta4 gene. *Mol. Cell. Biochem.* **272:** 97–105.

4. Filipowicz, A.W. & B.L. Horecker. 1983. In vitro synthesis of thymosin beta 4 encoded by rat spleen mRNA. *Proc. Natl. Acad. Sci. U. S. A.* **80:** 1811–1815.

5. Hannappel, E. 2010. Thymosin beta4 and its posttranslational modifications. *Ann. N. Y. Acad. Sci.* **1194:** 27–35.

6. Goldstein, A.L. *et al.* 2012. Thymosin beta4: a multifunctional regenerative peptide. Basic properties and clinical applications. *Expert Opin. Biol. Ther.* **12:** 37–51.

7. Hannappel, E. 2007. beta-Thymosins. *Ann. N. Y. Acad. Sci.* **1112:** 21–37.

8. Sanders, M.C., A.L. Goldstein & Y.L. Wang. 1992. Thymosin beta 4 (Fx peptide) is a potent regulator of actin polymerization in living cells. *Proc. Natl. Acad. Sci. U. S. A.* **89:** 4678–4682.

9. Sosne, G. *et al.* 2010. Biological activities of thymosin beta4 defined by active sites in short peptide sequences. *FASEB J.* **24:** 2144–2151.

10. Koutrafouri, V. *et al.* 2001. Effect of thymosin peptides on the chick chorioallantoic membrane angiogenesis model. *Biochim. Biophys. Acta.* **1568:** 60–66.

11. Malinda, K.M. *et al.* 1999. Thymosin beta4 accelerates wound healing. *J. Invest. Dermatol.* **113:** 364–368.

12. Badamchian, M. *et al.* 2003. Thymosin beta(4) reduces lethality and down-regulates inflammatory mediators in endotoxin-induced septic shock. *Int. Immunopharmacol.* **3:** 1225–1233.

13. Bock-Marquette, I. *et al.* 2004. Thymosin beta4 activates integrin-linked kinase and promotes cardiac cell migration, survival and cardiac repair. *Nature* **432:** 466–472.

14. Huff, T., C.S. Muller & E. Hannappel. 2007. Thymosin beta4 is not always the main beta-thymosin in mammalian platelets. *Ann. N. Y. Acad. Sci.* **1112:** 451–457.

15. Gerosa, C. *et al.* 2010. Thymosin beta-10 expression in developing human kidney. *J. Matern. Fetal. Neonatal Med.* **23**(Suppl. 3): 125–128.

16. Fanni, D. *et al.* 2011. Thymosin beta 10 expression in developing human salivary glands. *Early Hum. Dev.* **87:** 779–783.

17. Inzitari, R. *et al.* 2009. HPLC-ESI-MS analysis of oral human fluids reveals that gingival crevicular fluid is the main source of oral thymosins beta(4) and beta(10). *J. Sep. Sci.* **32:** 57–63.

18. Nemolato, S. *et al.* 2009. Thymosin beta(4) and beta(10) levels in pre-term newborn oral cavity and foetal salivary glands evidence a switch of secretion during foetal development. *PloS One* **4:** e5109.

19. Hay, E.D. 1995. An overview of epithelio-mesenchymal transformation. *Acta Anat.* **154:** 8–20.

20. Badamchian, M. *et al.* 2007. Identification and quantification of thymosin beta4 in human saliva and tears. *Ann. N. Y. Acad. Sci.* **1112:** 458–465.

21. Nemolato, S. *et al.* 2010. Thymosin beta 4 expression in normal skin, colon mucosa and in tumor infiltrating mast cells. *Eur. J. Histochem.* **54:** e3.

22. Larsson, L.I. & S. Holck. 2007. Occurrence of thymosin beta4 in human breast cancer cells and in other cell types of the tumor microenvironment. *Hum. Pathol.* **38:** 114–119.

23. Reti, R. *et al.* 2008. Thymosin beta4 is cytoprotective in human gingival fibroblasts. *Eur. J. Oral Sci.* **116:** 424–430.

24. Sosne, G. *et al.* 2002. Thymosin beta 4 promotes corneal wound healing and decreases inflammation in vivo following alkali injury. *Exp. Eye Res.* **74:** 293–299.

25. Smart, N., A. Rossdeutsch & P.R. Riley. 2007. Thymosin beta4 and angiogenesis: modes of action and therapeutic potential. *Angiogenesis* **10:** 229–241.

26. Goldstein, A.L. 2003. Thymosin beta4: a new molecular target for antitumor strategies. *J. Natl. Cancer Inst.* **95:** 1646–1647.

27. Sun, H.Q., K. Kwiatkowska & H.L. Yin. 1996. beta-Thymosins are not simple actin monomer buffering proteins. Insights from overexpression studies. *J. Biol. Chem.* **271:** 9223–9230.

28. Stossel, T.P., G. Fenteany & J.H. Hartwig. 2006. Cell surface actin remodeling. *J. Cell Sci.* **119:** 3261–3264.

29. Philp, D., A.L. Goldstein & H.K. Kleinman. 2004. Thymosin beta4 promotes angiogenesis, wound healing, and hair follicle development. *Mech. Ageing Dev.* **125:** 113–115.

30. Bonnet, D. *et al.* 1996. Thymosin beta4, inhibitor for normal hematopoietic progenitor cells. *Exp. Hematol.* **24:** 776–782.

31. Moon, H.S. *et al.* 2006. Zyxin is upregulated in the nucleus by thymosin beta4 in SiHa cells. *Exp. Cell Res.* **312:** 3425–3431.

32. Sun, H.Q., H.L. Yin. 2007. The beta-thymosin enigma. *Ann. N. Y. Acad. Sci.* **1112:** 45–55.

33. Kupatt, C. *et al.* 2005. Embryonic endothelial progenitor cells expressing a broad range of proangiogenic and remodeling factors enhance vascularization and tissue recovery in acute and chronic ischemia. *FASEB J.* **19:** 1576–1578.

34. Nemolato, S. *et al.* 2010. Different thymosin Beta 4 immunoreactivity in foetal and adult gastrointestinal tract. *PloS One.* **5:** e9111.

35. Nemolato, S. *et al.* 2011. Expression pattern of thymosin beta 4 in the adult human liver. *Eur. J. Histochem.* **55:** e25.

36. Nemolato, S. *et al.* 2012. Thymosin beta 4 in colorectal cancer is localized predominantly at the invasion front in tumor cells undergoing epithelial mesenchymal transition. *Cancer Biol. Ther.* **13:** 191–197.

37. Levayer, R. & T. Lecuit. 2008. Breaking down EMT. *Nat. Cell Biol.* **10:** 757–759.

38. Nakaya, Y. *et al.* 2008. RhoA and microtubule dynamics control cell-basement membrane interaction in EMT during gastrulation. *Nat. Cell Biol.* **10:** 765–775.

39. Siegrist, S.E. & C.Q. Doe. 2007. Microtubule-induced cortical cell polarity. *Genes Dev.* **21:** 483–496.

40. Thiery, J.P. 2003. Epithelial-mesenchymal transitions in development and pathologies. *Curr. Opin. Cell Biol.* **15:** 740–746.

41. Linker, C. *et al.* 2005. beta-Catenin-dependent Wnt signalling controls the epithelial organisation of somites through the activation of paraxis. *Development* **132:** 3895–3905.

42. Garber, K. 2008. Epithelial-to-mesenchymal transition is important to metastasis, but questions remain. *J. Natl. Cancer Inst.* **100:** 232–233.

43. Baum, B., J. Settleman & M.P. Quinlan. 2008. Transitions between epithelial and mesenchymal states in development and disease. *Semin. Cell Dev. Biol.* **19:** 294–308.

44. Tse, J.C. & R. Kalluri. 2007. Mechanisms of metastasis: epithelial-to-mesenchymal transition and contribution of tumor microenvironment. *J. Cell. Biochem.* **101:** 816–829.

45. Guarino, M., B. Rubino & G. Ballabio. 2007. The role of epithelial-mesenchymal transition in cancer pathology. *Pathology* **39:** 305–318.

46. Wang, Z.Y. *et al.* 2012. Evaluation of thymosin beta4 in the regulation of epithelial-mesenchymal transformation in urothelial carcinoma. *Urol. Oncol.* **30:** 167–176.

47. Wang, W.S. *et al.* 2003. Overexpression of the thymosin beta-4 gene is associated with malignant progression of SW480 colon cancer cells. *Oncogene* **22:** 3297–3306.

48. Huang, H.C. *et al.* 2007. Thymosin beta4 triggers an epithelial-mesenchymal transition in colorectal carcinoma by upregulating integrin-linked kinase. *Oncogene* **26:** 2781–2790.

49. Mannherz, H.G. & E. Hannappel. 2009. The beta-thymosins: intracellular and extracellular activities of a versatile actin binding protein family. *Cell Motil. Cytoskeleton* **66:** 839–851.

50. Safer, D., R. Golla & V.T. Nachmias. 1990. Isolation of a 5-kilodalton actin-sequestering peptide from human blood platelets. *Proc. Natl. Acad. Sci. U. S. A.* **87:** 2536–2540.

51. Weber, A. *et al.* 1992. Interaction of thymosin beta 4 with muscle and platelet actin: implications for actin sequestration in resting platelets. *Biochemistry* **31:** 6179–6185.

52. Sosne, G. *et al.* 2005. Thymosin-beta4 modulates corneal matrix metalloproteinase levels and polymorphonuclear cell infiltration after alkali injury. *Invest. Ophthalmol. Vis. Sci.* **46:** 2388–2395.

53. Xu, G.J. *et al.* 1982. Synthesis of thymosin beta 4 by peritoneal macrophages and adherent spleen cells. *Proc. Natl. Acad. Sci. U. S. A.* **79:** 4006–4009.

54. Kannan, L. *et al.* 2007. Identification and characterization of thymosin beta-4 in chicken macrophages using whole cell MALDI-TOF. *Ann. N. Y. Acad. Sci.* **1112:** 425–434.

55. Naylor, P.H. *et al.* 1984. Thymosin in the early diagnosis and treatment of high risk homosexuals and hemophiliacs with AIDS-like immune dysfunction. *Ann. N. Y. Acad. Sci.* **437:** 88–99.

56. Hannappel, E. & M. van Kampen. 1987. Determination of thymosin beta 4 in human blood cells and serum. *J. Chromatogr.* **397:** 279–285.

57. Cohen, M.V. & J.M. Downey. 2008. Adenosine: trigger and mediator of cardioprotection. *Basic Res. Cardiol.* **103:** 203–215.

58. Young, J.D. *et al.* 1999. Thymosin beta 4 sulfoxide is an anti-inflammatory agent generated by monocytes in the presence of glucocorticoids. *Nat. Med.* **5:** 1424–1427.

59. Nemolato, S. *et al.* 2010. Immunoreactivity of thymosin beta 4 in human foetal and adult genitourinary tract. *Eur. J. Histochem.* **54:** e43.

60. Maulbecker, C.C. & P. Gruss. 1993. The oncogenic potential of Pax genes. *EMBO J.* **12:** 2361–2367.

61. Ozcan, A. *et al.* 2011. PAX 8 expression in non-neoplastic tissues, primary tumors, and metastatic tumors: a comprehensive immunohistochemical study. *Mod. Pathol.* **24:** 751–764.

Ann. N.Y. Acad. Sci. ISSN 0077-8923

ANNALS OF THE NEW YORK ACADEMY OF SCIENCES

Issue: *Thymosins in Health and Disease*

Protective effects of thymosin β4 on carbon tetrachloride-induced acute hepatotoxicity in rats

Karina Reyes-Gordillo,[1,2,*] Ruchi Shah,[1,2,*] Jaime Arellanes-Robledo,[1,2] Marcos Rojkind,[1] and M. Raj Lakshman[1,2]

[1]Department of Biochemistry and Molecular Biology, School of Medicine and Health Sciences, The George Washington University Medical Center, Washington, DC. [2]Lipid Research Laboratory, VA Medical Center, Washington, DC

Address for correspondence: Karina Reyes-Gordillo, Ph.D., Lipid Research Laboratory (151-T), Room 1F-150, VA Medical Center, 50 Irving Street NW, Washington, DC 204222. karrygor@hotmail.com

Thymosin β4 (Tβ4) plays a role in fibrosis, inflammation, and in the reparative process of injured cells and tissues. Here, we discuss our preliminary work on the protective effect of Tβ4 on carbon tetrachloride (CCl_4)-induced acute hepatotoxicity. Our studies thus far indicate that Tβ4 can prevent necrosis, inflammatory infiltration, and upregulation of α1(and 2) collagen, α-SMA, PDGF-β receptor, and fibronectin mRNA expression; in addition, Tβ4 can prevent downregulation of PPARγ and upregulation of MECP2 mRNA levels in acute liver injury. Our initial work therefore indicates that Tβ4 can prevent the alteration of markers of hepatic stellate cell transdifferentiation, which suggests that Tβ4 could maintain the quiescent phenotypic state of hepatic stellate cells in the rat livers by restoring PPARγ and downregulating MeCP2 expression levels. More specifically, these preliminary studies suggest that Tβ4 might be an effective anti-inflammatory and antifibrotic drug for the treatment of liver fibrogenesis.

Keywords: carbon tetrachloride; hepatotoxocity; collagen; PPARγ; thymosin β4

Introduction

Liver injury typically induces hepatocyte necrosis and apoptosis regardless of its causes. Necrosis engages classic inflammatory and fibrogenic signals.[1] Liver damage can be caused by viral infection, autoimmune disorders, ischemia, and several xenobiotics, including: drugs, alcohol, or toxins.[2] The carbon tetrachloride (CCl_4)-induced acute liver injury model is widely used to investigate the mechanisms of liver damage and regeneration.[3] Treatment with CCl_4, a known hepatotoxin, stimulates experimental acute liver failure through free radical-mediated peroxide injuries.[4] This treatment is accompanied by extensive necrosis and inflammation.[5] Even though during acute liver damage there is no fibrosis, there is activation of hepatic stellate cells.[6] Hepatic stellate cells are the main fibrogenic cells of the injured liver. In their normal (quies-

cent) stage they mainly produce an extracellular matrix present in basement membranes such as type IV collagen.[7,8] They store vitamin A and triglycerides and express regulators of the adipocyte phenotype such as peroxisome proliferator-activated receptor (PPARγ), sterol regulatory element binding-protein1 (SREBP-1c), and methyl-CpG binding protein 2 (MeCP2) among others.[9–12]

In the fibrotic liver, hepatic stellate cells undergo transdifferentiation from lipid-storing pericytes to myofibroblastic cells (Fig. 1). This activation requires coordinated changes in activity of several growth factors such as the platelet-derived growth factor (PDGF) and the transforming growth factor β1 (TGF-β1).[13,14] Specifically, PDGF is the most potent proliferative cytokine acting on hepatic stellate cells.[15] Activated hepatic stellate cells show significant alterations in gene expression, where expressions of PPARγ and SREBP-1c are downregulated,[11,15,16] while expression of MeCP2 is upregulated.[12] Hepatic stellate cells lose the retinoid-binding proteins and their vitamin A

---

*These authors contributed equally to this work.

doi: 10.1111/j.1749-6632.2012.06728.x

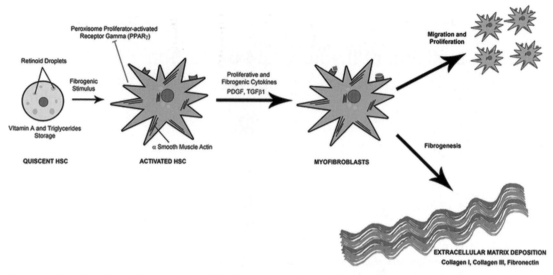

Figure 1. Hepatic stellate cells are perisinusoidal cells of the liver that are involved in the production of extracellular matrix components present in the space of Disse and in regulating blood flow. Hepatic stellate cells are vitamin A–storing cells that undergo phenotypic transdifferentiation characterized as "myofibroblastic activation" during liver fibrogenesis. Activated hepatic stellate cells lose the vitamin A stores and express cytokine receptors like PDGF-β receptor. Moreover, they acquire a contractile cytoskeleton and express alpha smooth muscle actin (α-SMA) and skeletal myosins, which are the markers of HSC transdifferentiation. The activated myofibroblasts then, migrate and proliferate to the site of injury and form a fibrous scar. In addition, they also deposit extracellular matrix proteins such as collagen I, III, IV, and fibronectin, among others. This activation process involves morphological and biochemical changes. Peroxisome proliferator-activated receptor gamma (PPARγ) is one such factor whose activity is decreased in activated hepatic stellate cells. PPARγ ligands suppress several markers of hepatic stellate cell activation such as expression of collagen and α-SMA, cell proliferation, and migration. These findings support the role of PPARγ in reversion and/or prevention of activated hepatic stellate cell toward their quiescent state.

stores.[17] The activated hepatic stellate cells are proliferative, proinflammatory and fibrogenic with the inducedability to synthesize and deposit large amounts of extracellular matrix proteins.[7,8] Also, activated hepatic stellate cells overexpress genes that confer the myofibroblastic phenotype such as collagens I and III, fibronectin and *de novo* synthesis of α-smooth muscle actin (α-SMA).[7,8,17] Thus, a better understanding of the mechanism underlying hepatic stellate cell transdifferentiation is pivotal for identifying molecular targets required for the development of new therapeutic treatments for liver damage.

Thymosin β4 (Tβ4) is a 43 amino acid polypeptide that was initially isolated from calf thymus.[18] It is a component of a family of approximately 15 members with a highly conserved amino acid sequence.[19] Tβ4 has been shown to prevent inflammation and fibrosis, promoting healing in the eye, skin, and heart;[20–23] in the eye, it promotes corneal reepithelialization after skin injury. Tβ4 also inhibits a strong inflammatory component

that occurs after injury with sodium hydroxide.[19,20] Overall, it prevents inflammation by blocking the secretion of inflammatory cytokines and suppressing the activation of nuclear factor-κB.[24] In the heart, Tβ4 prevents the formation of scar tissue after a myocardial infarction by enhancing the survival of myocardial tissue and endothelial cells, thus, sustaining cardiac function and preventing scar formation.[22,23] Recently, it has been shown that Tβ4 inhibits the appearance of myofibroblasts in a model system of wound healing.[25] Our previous studies have revealed that rat hepatic stellate cell clones derived from cirrhotic rat liver express Tβ4;[26] moreover, the addition of Tβ4 to hepatic stellate cells/myofibroblasts cultures inhibits PDGF-β receptor expression and prevents binding of protein kinase B (AKT) to actin and its phosphorylation by phosphoinositide dependent kinase-1 (PDK1) and mammalian target of rapamycin (mTOR).[27] Based on these findings, we believe that Tβ4 could have therapeutic properties to prevent liver injury.

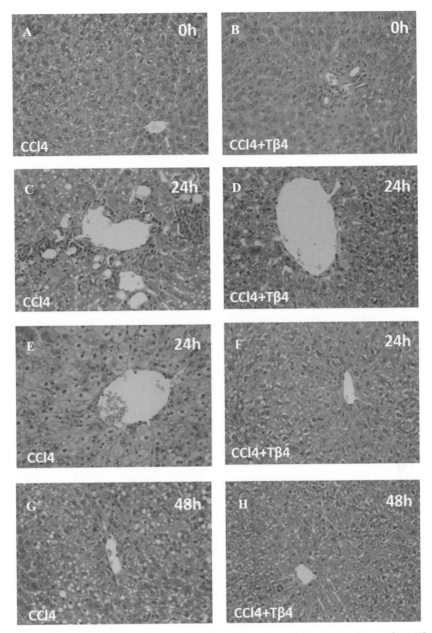

Figure 2. Hematoxylin and eosin staining of liver sections from (A) rats treated with CCl₄ with or without Tβ4 at (A and B) 0 h, (C–F) 24 h, and (G and H) 48 h, respectively. Panels C and E shows the presence of portal inflammation, centrizonal necrosis, and distortion of liver around portal triads, vacuole generation, and microvascular steatosis, 24 h after CCl₄ treatment. As shown panels D and F, Tβ4 prevented histological changes in CCl₄-treated rat livers. These data are representative of work in progress.

In an ongoing study, we have been investigating the potential of Tβ4 to inhibit acute liver damage induced by CCl₄ in an *in vivo* model. Our results to date indicate that Tβ4 by preventing the hallmarks of hepatic stellate cells activation and liver fibrosis significantly reduces CCl₄-induced inflammatory infiltration by restoring the expression and the level of the adipogenic transcription factor PPARγ, while inhibiting the upregulation of MeCP2.

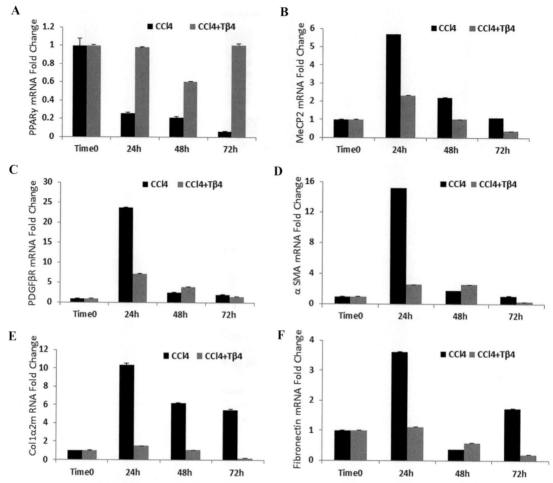

Figure 3. Quantitative RT-PCR analysis of (A) PPARγ, (B) MeCP2, (C) PDGF-β receptor, (D) α-SMA, (E) Collagen 1α2, (F) Fibronectin mRNA. Total RNA was extracted from whole livers of rats treated with either CCl₄ or CCl₄ plus Tβ4 at 1 mg/kg body weight in various time points indicated in the figure. All the values are means of triplicate experiments and were corrected with GAPDH mRNA expression. These data are representative of work in progress.

Tβ4 protects against CCl₄-induced acute liver damage

We first examined the effect of Tβ4 on acute liver damage by a histological approach (Fig. 2); a representative liver tissue section from control rats (time 0) is shown. In rats that received CCl₄ alone, liver histology at 24 h after CCl₄ exposure showed extensive hepatocellular damage, as evidenced by the presence of portal inflammation, centrizonal necrosis, and distortion of the liver around portal triads. Vacuole generation and microvascular steatosis were also observed (Fig. 2C and 2E). These histopathological changes were ameliorated by Tβ4 treatment (Fig. 2D and 2F). No significant differences were

found after 48 h of CCl₄ administration. Together, these results indicate that Tβ4 has the potential to inhibit the acute CCl₄-induced liver injury.

CCl₄-induced activation of hepatic stellate cells

During the initiation of CCl₄-induced acute liver damage, hepatic stellate cells differentiate into myofibroblast-like cells.[7,8] Myofibroblast transdifferentiation is characterized by changes in the expression of the following genes that, together, generate the myofibroblast phenotype.[7] Specifically, MeCP2 operates as a powerful epigenetic repressor of gene transcription.[12] In addition, transcriptional

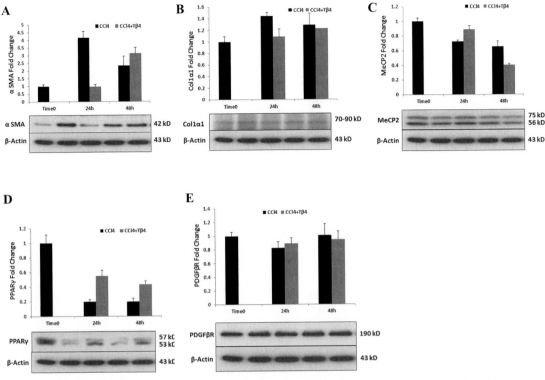

Figure 4. Western blot analysis performed with total proteins extracted from whole rat livers treated with CCl₄ in the presence or absence of Tβ4. The protein expression of (A) α-SMA, (B) collagen 1α1, and (C) MeCP2 was decreased after 24 h of CCl₄ treatment; Tβ4 restored the protein expression to normal levels similar to that of the control. Values are means of triplicate experiments ±SE and were corrected for possible differences in loading after reprobing with an antibody to β actin. These data are representative of work in progress.

silencing of the PPARγ gene is required for conversion of hepatic stellate cells into myofibroblasts. PPARγ expression is associated with the adipogenic features of quiescent hepatic stellate cells and must be silenced in these cells to adopt myofibroblast characteristics.[14]

To determine whether Tβ4 prevents adipogenic transcriptional regulation during hepatic stellate cells transdifferentiation, we are evaluating its effects on CCl₄-induced liver injury by measuring MeCP2 and PPARγ mRNA expression using a qPCR assay. Our work thus far has shown that 24 h after CCl₄ administration, PPARγ mRNA level is lower in the CCl₄ group than in the control group and that Tβ4 treatment can prevent this effect (Fig. 3A). Furthermore, CCl₄ administration produces an increase in the MeCP2 mRNA level that can be significantly decreased by Tβ4 (Fig. 3B).

Expression of the PDGF-β receptor in hepatic stellate cells also plays a key role in their activation

and transformation into myofibroblasts.[13] We have therefore also examined the effect of Tβ4 on mRNA expression of PDGF-β receptor using qPCR analysis. As illustrated in Figure 3C, initial studies have shown that while levels of PDGF-β receptor increase after 24 h in the CCl₄-treated group, Tβ4-treatment can significantly inhibit the expression of this receptor; similar results are seen for an established marker of hepatic stellate cells activation, namely α-SMA (Fig. 3D) and for mRNAs pertaining to deposition of extracellular matrix, such as collagen (Fig. 3E) and fibronectin (Fig. 3F). These preliminary data support our working hypothesis that Tβ4 protects against CCl₄-induced liver damage.

Another method we have used to investigate the effect of Tβ4 on acute liver damage is by measuring the production of several proteins, including PDGF-β receptor, α-SMA, type I collagen, PPARγ, and MeCP2 by Western blot analysis. An example of this type of analysis is shown in Figure 4,

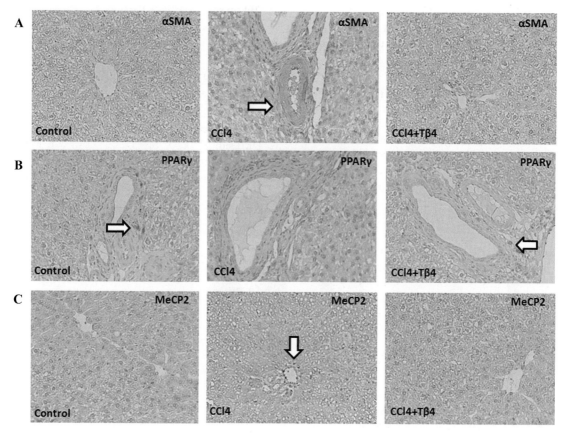

Figure 5. Immunostaining of liver sections from CCl₄-treated rats with or without Tβ4 after 24 h. Panel A shows α-SMA positive staining in perisinusoidal spaces and in fibrous septa of liver sections of CCl₄-treated rats. Tβ4 prevents the α-SMA positive staining. Panel B shows PPARγ-positive cells in intralobular areas; however, the staining is less evident in liver sections from CCl₄-treated rats. Tβ4 restores the levels of PPARγ after 24 h. As shown in panel C, MeCP2 positive staining is observed in periportal areas while in Tβ4-treated tissue the staining is imperceptible after 24 h. These data are representative of work in progress.

in which Tβ4 treatment blocks CCl₄-induced increases in the expressions of these proteins. Similarly, we have used immunostaining for α-SMA, MeCP2, and PPARγ in liver tissue sections after CCl₄ alone or after CCl₄ plus Tβ4 treatment (Fig. 5). Taken together, our work thus far indicates that Tβ4 inhibits activation of hepatic stellate cells by restoring the expression of adipogenic transcription factors PPARγ and MeCP2, while suppressing hepatic stellate cells activation markers.

Discussion

Acute and chronic liver diseases constitute a global concern. At present, there is no approved therapy to treat these diseases even in the developed world. Therefore, research for therapeutically effective agents for the treatment of liver diseases is highly relevant and requires urgent attention. This paper describes work in progress that supports the conclusion that Tβ4 can, at least in a rat model, attenuate CCl₄-induced hepatotoxicity. The mechanism for this protective effect could be related to the fact that Tβ4 efficiently inhibits hepatic stellate cell activation *in vivo*.

CCl₄ acute exposure is probably the most widely used model for screening the hepatoprotective activity of drugs. CCl₄ is a potent hepatotoxin, and a single exposure can lead rapidly to severe hepatic necrosis, steatosis, and portal inflammation.[28,29] Our studies indicate that Tβ4 preserves the hepatocellular membrane and suppresses CCl₄-induced acute liver injury by reducing the inflammatory infiltration, necrosis, and microvascular steatosis observed during histological analysis.

According to current concepts, upon liver injury, hepatic stellate cells proliferate and differentiate into myofibroblast-like cells. The activated hepatic stellate cells undergo continuous proliferation and express activation markers such as α-SMA and produce large amounts of extracellular matrix proteins, including type I collagen.[7,8] One of the key events in the activation of hepatic stellate cells is the expression of the PDGF-β receptor.[13] In our experimental system, increased expressions of α-SMA, PDGF-β receptor, and collagen type I expression are consistent with the conclusion that hepatic stellate cells are activated by CCl$_4$-induced liver injury and that Tβ4 treatment can suppress this activation. This conclusion is supported by our previous study[27] in cultured hepatic stellate cells showing that *in vitro* Tβ4 prevents hepatic stellate cells/myofibroblast transdifferentiation, proliferation, and migration, and that, in part, the effect of Tβ4 is mediated by its capacity to bind to actin and AKT binding site. Additionally, we showed that binding to actin was required by PDK1 and by mTOR to phosphorylate AKT at threonine 308 and serine 473, respectively.[27] Thus, our previously published and current work in progress, suggest that Tβ4 is antifibrogenic by preventing and/or reversing the activation of hepatic stellate cells.

As described above, PPARγ is epigenetically repressed by induction of MeCP2 and a shift from adipogenic to myogenic mode characterizes hepatic stellate cell transdifferentiation. Mann *et al.*[12] recently described that the loss of expression of the master adipogenic regulator PPARγ plays a central role in this shift. Therefore, the restoration of PPARγ expression and/or other adipogenic transcription factors reverses myofibroblastic hepatic stellate cells to differentiated cells.[12,30] Similar data have been reported by Milam *et al.*,[31] who showed that PPARγ regulates adipogenesis, inflammatory responses, and exerts potent antifibrotic effects *in vitro* and *in vivo*. They observed that PPARγ activation abrogated collagen synthesis, cell migration, and myofibroblast transdifferentiation induced by TGF-β in lung fibroblasts.[31]

Our work in progress supports the conclusion that the antifibrotic properties of Tβ4 is related to the restoration of adipogenic transcription factor PPARγ expression and downregulation of MeCP2. And while our work to date shows a protective effect of Tβ4 on acute liver damage in a short-term study, further research is required to assess its effect on reverting already established liver fibrosis.

Acknowledgments

We would thank Dr. Allan L. Goldstein, for his encouragement and advice and Dr. Hynda K. Kleinman, for her advice. We also express our gratitude to Dr. Gadi Spira for his advice and Ethan Matz for correcting the manuscript. The synthetic Tβ4 used for this work was a kind gift from RegeneRx Biopharmaceuticals, Inc. This work was supported with NIAAA Grant RO1 10541 (MRL).

Conflict of interest

The authors declare no conflict of interest.

References

1. Cohen-Naftaly, M. & S.L. Friedman. 2011. Current status of novel antifibrotic therapies in patients with chronic liver disease. *Therap. Adv. Gastroenterol.* **4:** 391–417.

2. Kim, H.Y., J.K. Kim, J.H. Choi, *et al.* 2010. Hepatoprotective effect of pinoresinol on carbon tetrachloride-induced hepatic damage in mice. *J. Pharmacol. Sci.* **112:** 105–112.

3. Weber, L.W., M. Boll & A. Stampfl. 2003. Hepatotoxicity and mechanism of action of haloalkanes: carbon tetrachloride as a toxicological model. *Crit. Rev. Toxicol.* **33:** 105–136.

4. Muriel, P. 1997. Peroxidation of lipids and liver damage. In *Oxidants and Free Radicals*. S. Baskin and H. Salem, Eds.: 237–257. Taylor & Francis. Washington.

5. Reyes-Gordillo, K., J. Segovia, M. Shibayama, *et al.* 2007. Curcumin protects against acute liver damage in the rat by inhibiting NF-kappaB, proinflammatory cytokines production and oxidative stress. *Biochim. Biophys. Acta* **1770:** 989–996.

6. Hellerbrand, C., B. Stefanovic, F. Giordano, *et al.* 1999. The role of TGF beta1 in initiating hepatic stellate cell activation in vivo. *J. Hepatol.* **30:** 77–87.

7. Safadi, R. & S.L. Friedman. 2002. Hepatic fibrosis—role of hepatic stellate cell activation. *MedGenMed.* **4:** 27.

8. Rojkind, M., Reyes-Gordillo, K. 2009. Hepatic stellate cells. In *The liver Biology and Pathobiology*. I.M.W.A.W. Arias, J.L. Boyer, D.E. Cohen, D.A. Shafritz & N. Fausto, H.J. Alter, Eds.: 407–432. Willey Blackwell. Oxford, UK.

9. Lee, T.F., K.M. Mak, O. Rackovsky, *et al.* 2010. Down-regulation of hepatic stellate cell activation by retinol and palmitate mediated by adipose differentiation-related protein (ADRP). *J. Cell Physiol.* **223:** 648–657.

10. Shafiei, M.S., S. Shetty, P.E. Scherer, *et al.* 2011. Adiponectin regulation of stellate cell activation via PPARgamma-dependent and -independent mechanisms. *Am. J. Pathol.* **178:** 2690–2699.

11. Tsukamoto, H., N.L. Zhu, K. Asahina, *et al.* 2011. Epigenetic cell fate regulation of hepatic stellate cells. *Hepatol. Res.* **41:** 675–682.

12. Mann, J., F. Oakley, F. Akiboye, *et al.* 2007. Regulation of myofibroblast transdifferentiation by DNA methylation and MeCP2: implications for wound healing and fibrogenesis. *Cell Death Differ.* **14:** 275–285.

13. Brenner, D.A. 2009. Molecular pathogenesis of liver fibrosis. *Trans. Am. Clin. Climatol. Assoc.* **120:** 361–368.

14. Hernandez-Gea, V. & S.L. Friedman. 2011. Pathogenesis of liver fibrosis. *Annu. Rev. Pathol.* **6:** 425–456.

15. She, H., S. Xiong, S. Hazra, *et al.* 2005. Adipogenic transcriptional regulation of hepatic stellate cells. *J. Biol. Chem.* **280:** 4959–4967.

16. Hazra, S., T. Miyahara, R.A. Rippe, *et al.* 2004. PPAR gamma and hepatic stellate cells. *Comp. Hepatol.* **3**(Suppl 1)**:** S7.

17. Mann, J. & D.A. Mann. 2009. Transcriptional regulation of hepatic stellate cells. *Adv. Drug Deliv. Rev.* **61:** 497–512.

18. Badamchian, M., A.A. Damavandy, H. Damavandy, *et al.* 2007. Identification and quantification of thymosin beta4 in human saliva and tears. *Ann. N. Y. Acad. Sci.* **1112:** 458–465.

19. Goldstein, A.L., E. Hannappel & H.K. Kleinman. 2005. Thymosin beta4: actin-sequestering protein moonlights to repair injured tissues. *Trends Mol. Med.* **11:** 421–429.

20. Dunn, S.P., D.G. Heidemann, C.Y. Chow, *et al.* 2010. Treatment of chronic nonhealing neurotrophic corneal epithelial defects with thymosin beta 4. *Arch Ophthalmol.* **128:** 636–638.

21. Philp, D., A.L. Goldstein & H.K. Kleinman. 2004. Thymosin beta4 promotes angiogenesis, wound healing, and hair follicle development. *Mech. Ageing Dev.* **125:** 113–115.

22. Bock-Marquette, I., A. Saxena, M.D. White, *et al.* 2004. Thymosin beta4 activates integrin-linked kinase and promotes cardiac cell migration, survival and cardiac repair. *Nature* **432:** 466–472.

23. Smart, N., C.A. Risebro, A.A. Melville, *et al.* 2007. Thymosin beta4 induces adult epicardial progenitor mobilization and neovascularization. *Nature* **445:** 177–182.

24. Sosne, G., P. Qiu, M. Kurpakus-Wheater, *et al.* 2010. Thymosin beta4 and corneal wound healing: visions of the future. *Ann. N. Y. Acad. Sci.* **1194:** 190–198.

25. Ehrlich, H.P. & S.W. Hazard, 3rd. 2010. Thymosin beta4 enhances repair by organizing connective tissue and preventing the appearance of myofibroblasts. *Ann. N. Y. Acad. Sci.* **1194:** 118–124.

26. Barnaeva, E., A. Nadezhda, E. Hannappel, *et al.* 2007. Thymosin beta4 upregulates the expression of hepatocyte growth factor and downregulates the expression of PDGF-beta receptor in human hepatic stellate cells. *Ann. N. Y. Acad. Sci.* **1112:** 154–160.

27. Reyes-Gordillo, K., R. Shah, A. Popratiloff, *et al.* 2011. Thymosin-beta4 (Tbeta4) blunts PDGF-dependent phosphorylation and binding of AKT to actin in hepatic stellate cells. *Am. J. Pathol.* **178:** 2100–2108.

28. Manibusan, M.K., M. Odin & D.A. Eastmond. 2007. Postulated carbon tetrachloride mode of action: a review. *J. Environ. Sci. Health C. Environ. Carcinog. Ecotoxicol. Rev.* **25:** 185–209.

29. Recknagel, R.O., E.A. Glende, Jr., J.A. Dolak, *et al.* 1989. Mechanisms of carbon tetrachloride toxicity. *Pharmacol. Ther.* **43:** 139–154.

30. Mann, J., D.C. Chu, A. Maxwell, *et al.* 2010. MeCP2 controls an epigenetic pathway that promotes myofibroblast transdifferentiation and fibrosis. *Gastroenterology* **138:** 705–714, 714 e1–4.

31. Milam, J.E., V.G. Keshamouni, *et al.* 2008. PPAR-gamma agonists inhibit profibrotic phenotypes in human lung fibroblasts and bleomycin-induced pulmonary fibrosis. *Am J Physiol Lung Cell Mol Physiol.* **294:** L891–L901.

Ann. N.Y. Acad. Sci. ISSN 0077-8923

Protective effects of thymosin β4 in a mouse model of lung fibrosis

Enrico Conte,[1] Tiziana Genovese,[2] Elisa Gili,[1] Emanuela Esposito,[2] Maria Iemmolo,[1] Mary Fruciano,[1] Evelina Fagone,[1] Maria Provvidenza Pistorio,[1] Nunzio Crimi,[1] Salvatore Cuzzocrea,[2] and Carlo Vancheri[1]

[1]Department of Clinical and Molecular Biomedicine, University of Catania, Italy. [2]Department of Clinical and Experimental Medicine and Pharmacology, School of Medicine, University of Messina, Italy

Address for correspondence: Enrico Conte, University of Catania—Clinical and Molecular Biomedicine, A. O. Policlinico via S. Sofia 78, Catania, Sicily 95123, Italy. econte@unict.it

Thymosin β4 (Tβ4) has been found to have several biological activities related to antiscarring and reduced fibrosis. For example, the anti-inflammatory properties of Tβ4 and its splice variant have been shown in the eye and skin. Moreover, Tβ4 treatment prevents profibrotic gene expression in cardiac and in hepatic cells *in vitro* and *in vivo*. In a recent study on scleroderma patients it was hypothesized that Tβ4 may exert a protective effect during human lung injury. In an ongoing study, we have explored the putative Tβ4 protective role in the lung context by utilizing a well-known *in vivo* model. We have observed significant protective effects of Tβ4 on bleomycin-induced lung damage, the main outcomes being the halting of the inflammatory process and a substantial reduction of histological evidence of lung injury.

Keywords: thymosin β4; lung; inflammation; fibrosis; mouse; bleomycin

Common pathologic features of interstitial lung diseases, including idiopathic pulmonary fibrosis (IPF), comprise fibrosis of the interstitium involving increased collagen content in the matrix, elastic, and smooth muscle proliferation, alveolar architectural remodeling, and chronic inflammation.[1] IPF is a relentless disease characterized by alveolar epithelial cell injury and hyperplasia, inflammatory cell accumulation, fibroblast hyperplasia, deposition of extracellular matrix, scar formation, and the consequent loss of respiratory function.[2] No approved, effective pharmacologic treatment is available for IPF, except pirfenidone, which may delay clinical worsening. Novel strategies are required for its management of, as well as a better understanding of, the molecular mechanisms underlying the pathogenesis and progression of this disease. No animal model exactly reproduces the evolution of the fibrotic process, nevertheless, the bleomycin model of pulmonary fibrosis in mice is the best characterized and recapitulates main aspects of the pathology.

Thymosin β4 (Tβ4) belongs to a broad family of bioactive peptides that are highly conserved across species and represent the major G-actin–sequestering proteins in cells. This small ubiquitous protein containing 43 amino acids exerts a well-established intracellular function in regulating monomeric actin and thus controlling its polymerization.[3,4] Tβ4 also has broad biological activities and pleiotropic effects on different types of cells. These functions are distinguishable and significant in wound healing and in inflammation, such as promoting cell motility and downregulating inflammatory cytokines, respectively, while other properties are important to both processes such as promoting cell adhesion and inhibiting apoptosis.[3]

Tβ4 anti-inflammatory properties

Previous observations have demonstrated a significant anti-inflammatory outcome of exogenous Tβ4 in the eye.[5,6] In particular, in a mouse alkali burn model (marked by a massive neutrophil infiltration and inflammatory response characterized by overexpression of proinflammatory cytokines/chemokines and matrix-metalloproteinases, much more extreme than that seen in the setting of scrape injury), Tβ4 was shown to inhibit neutrophil infiltration, decreasing corneal KC (CXCL1) and macrophage inflammatory protein (MIP)-2 chemokine expression, and to modulate corneal matrix metalloproteinase levels.[7] Moreover, it was reported that Tβ4 sulfoxide, which results from oxidation of the methionine residue at position 6 of Tβ4, is generated by monocytes in the presence of glucocorticoids and may act as a signal to inhibit the inflammatory response.[8] *In vitro* Tβ4 sulfoxide inhibited PMN chemotaxis to the bacterial chemoattractant, *N*-formyl-methionyl-leucyl-phenylalanine (fMLP), whereas *in vivo* it was a potent inhibitor of carrageenan-induced edema in the mouse paw. On the other hand, in an *in vivo* skin model, the Tβ4 splice-variant derived from lymphocytes was shown to exert broader and stronger antiinflammatory activities, including the inhibition of neutrophilic infiltration, as assessed by three different strategies: footpad λ-carrageenan injection, irritant contact dermatitis to 12-*O*-tetradecaneylphorbol 13-acetate, and allergic contact dermatitis to 2,4-dinitrofluorobenzene.[9]

In a separate study, it was shown that Tβ4 substantially lowered levels of circulating inflammatory cytokines and reactive mediators following LPS administration in vivo.[10] The same paper demonstrated that Tβ4 levels quickly disappeared in the blood following LPS administration, in both humans and rodents. It is also significantly decreased in patients with septic shock, thus suggesting that it may be involved in early events determining the activation of the inflammatory cascade and clinical consequences of sepsis.

In recent ongoing studies, we have investigated for the first time the putative protective role of Tβ4 in the bleomycin model of lung inflammation/fibrosis. In ongoing work, we are evaluating the effects of Tβ4 cotreatment (intraperitoneally, 6 mg/kg every three days) after seven days of bleomycin-induced lung injury when the inflammatory phase is completely displayed and an ongoing fibrotic process is evaluable. We have compared leukocytes number (in particular PMN cells) in bronchoalveolar lavage fluids as well as myeloperoxidase (MPO) activity in the lung tissues and levels of edema. MPO employs

hydrogen peroxide derived by dismutation of superoxide to produce hypochlorous acid, a compound with relevant antibacterial properties yet superoxide can react with nitric oxide to generate highly reactive metabolites such as peroxynitrite. This compound is able to oxidize proteins, resulting in direct nitration of tyrosine residues (n-Tyr). Protein structure and function can be subsequently altered and enzymatic activity affected. Proteins containing n-Tyr residues have been previously detected in the lung of bleomycin treated animals. As shown in Figure 1, the huge increase in n-Tyr staining of lung sections of bleomycin animals after one week of treatment was almost completely abolished in the Tβ4 cotreated mice. Moreover, based on additional parameters, the bleomycin-induced inflammatory process after one week of treatment was substantially inhibited in Tβ4 cotreated mice.

Mechanisms of Tβ4 anti-inflammatory activity

The mechanisms by which Tβ4 regulates the inflammatory response following injuries are poorly understood. Even though there are many inflammatory signaling pathways, the transcription factor NF-κB, regulating the expression of a wide variety of inflammatory genes, is believed to play a pivotal role in the inflammatory process.[11,12] A

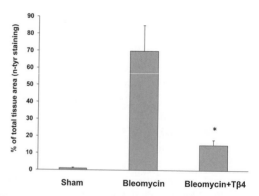

Figure 1. Effect of Tβ4 on nitration of tyrosine residues in the lung after one week of bleomycin treatment. Nitration of tyrosine residues (n-tyr) accounts for the oxidative damage, as a consequence of massive myeloperoxidase activity. The presence of n-tyr residues, detected by immunohistochemistry, was quantified (by image software) in microphotographs of random chosen fields of lung sections and reported in the graph as percentage of total area. The huge increase in n-tyr content observed in bleomycin-treated mice was significantly inhibited in Tβ4 cotreated mice. These data represent work in progress.

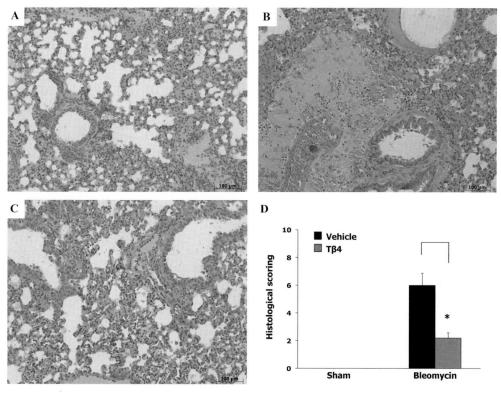

Figure 2. Effect of Tβ4 on bleomycin-induced lung injury and fibrosis. Representative microphotographs (150×) of paraffin-embedded formalin-fixed lung tissue slices conventionally stained by hematoxylin. The normal lung architecture observable in sham operated animals (A), appears substantially damaged in bleomycin-treated mice (B), whereas a quite complete recovery results in Tβ4-cotreated animals (C). The semiquantitative lung fibrosis scoring (D) was assessed according by the Ashcroft scale. Criteria for grading were as follows: grade 0, normal lung; grade 1, minimal fibrous thickening of alveolar or bronchiolar walls; grade 3, moderate thickening of walls without obvious damage to lung architecture; grade 5, increased fibrosis with definite damage to lung structure and formation of fibrous bands or small fibrous masses; grade 7, severe distortion of structure and large fibrous areas; grade 8, total fibrous obliteration of fields. Grades 2, 4, and 6 were used as intermediate pictures between the aforementioned criteria. The asterisk indicates a $P < 0.05$. These data represent work in progress.

recent paper[13] has shown evidence that in corneal and conjunctival epithelial cells Tβ4 directly targets the NF-κB RelA/p65 subunit, blocking RelA/p65 nuclear translocation and binding to the cognate κB site in the proximal region of the IL-8 gene promoter. Furthermore, enforced expression of Tβ4 interferes with TNF-α–mediated NF-κB activation, as well as downstream IL-8 gene transcription. Importantly, those activities were shown independent of the G-actin–binding properties of Tβ4. Moreover, in a mouse myocardial infarction model, Tβ4 has been shown to substantially inhibit NF-κB activation, resulting in a significant improvement of cardiac function.[14] Based on these findings, it is reasonable to speculate that some Tβ4 anti-inflammatory properties lie in mediating NF-κB signaling pathways. Another mechanism implies re-

active oxygen species (ROS). On the other hand, also by activating NF-κB, ROS play a crucial role in inflammation when the balance between and endogenous antioxidants preserved in healthy tissues is broken and ROS prevail. Importantly, studies on human cornea epithelial cells have shown that exogenous Tβ4 is able to stimulate the expression of manganese superoxide dismutase (SOD) and copper/zinc SOD as well as upregulate both the transcription and translation levels of catalase in the presence of exogenous H_2O_2 and protect against oxidative damage.[15,16]

Moreover, it has been demonstrated that Tβ4 treatment selectively targets and upregulates catalase and Cu/Zn-SOD also in cardiac fibroblasts.[17] Since the inflammatory effects of bleomycin are known to be dependent on ROS generation, Tβ4 lung protection from bleomycin effects in our experimental setting could be based on such Tβ4-induced antioxidative responses.

Tβ4 antifibrotic properties

It has shown that in a mouse model of acute myocardial infarction exogenous Tβ4 significantly reduces interstitial cardiac fibrosis and aids in cardiac function and repair, by enhancing the level of PINCH-1-ILK-α-parvin (PIP) components and Akt activation, while substantially suppressing collagen expression, the hallmark of cardiac fibrosis.[14] Previous studies demonstrated that Tβ4 plays an important role in healing of hypoxic injury in the heart by preventing apoptotic cell death of cardiomyocytes and reducing scarring.[18] Moreover, Tβ4 was shown to upregulate the expression of hepatocyte growth factor and downregulate the expression of platelet-derived growth factor-beta receptor in a model of liver fibrosis suggesting the antifibrotic potential of Tβ4 in the liver.[19] Similarly, the peptide derived from the Tβ4 aminoterminal end, *N*-acetyl-seryl-aspartyl-lysyl-proline (Ac-SDKP), was shown to mediates the antifibrogenic effects of angiotensin converting enzyme inhibitors on heart and kidney fibrosis,[20] and there is convincing evidence that chronic treatment with Ac-SDKP reduces collagen deposition in the heart and kidneys of hypertensive rats and rats with myocardial infarction.[21,22] Moreover, in a previously published paper, rat neonatal cardiac fibroblasts were used to study the effects of Tβ4 under oxidative stress, and it was shown that Tβ4 treatment prevented profibrotic gene expression in this *in vitro* setting.[17] In a recent paper, Tβ4 was shown to be present at high concentration in the BALF of scleroderma patients with lung involvement and, since lower levels of Tβ4 in BALF were associated with interstitial lung disease progression, it has been hypothesized that Tβ4 may exert a cytoprotective effect during lung injury in this disease.[23]

In our *in vivo* experimental model, bleomycin-induced lung damage and fibrosis are evaluated by histological examination and the severity of fibrosis is semiquantitatively assessed according to the method of Ashcroft *et al.*[24] As shown in Figure 2, our work in progress shows that seven days from drug delivery the initial bleomycin-induced fibrosis (as evidenced by an increase in extracellular matrix deposition, thickening of alveolar and bronchiolar walls, damage to lung architecture by fibrous band) is significantly inhibited by cotreatment with exogenous Tβ4, administered intraperitoneally. Based on these preliminary studies, and multiple studies in animal models of fibrotic pathologies and data from human studies, we believe that there is a sound rationale for expecting Tβ4 to play an important role in pathologies such as idiopathic pulmonary fibrosis.

Tβ4 is a multifunctional, regenerative, and antifibrotic molecule. It has a potent effect on reducing inflammation and has been shown to prevent scarring and fibrosis in many different animal models of human pathologies. These preliminary findings provide the foundation for future studies on its role in fibrosis and scarring and demonstrate a rationale for its use in the clinic.

Conflicts of interest

The authors declare no conflicts of interest.

References

1. Green, F.H. 2002. Overview of pulmonary fibrosis. *Chest* **122:** 334S–339S.
2. American Thoracic Society/European Respiratory Society International Multidisciplinary Consensus Classification of the Idiopathic Interstitial Pneumonias. 2002. This joint statement of the American Thoracic Society (ATS), and the European Respiratory Society (ERS) was adopted by the ATS board of directors, June 2001 and by the ERS Executive Committee, June 2001. *Am. J. Respir. Crit Care Med.* **165:** 277–304.
3. Goldstein, A.L., E. Hannappel & H.K. Kleinman. 2005. Thymosin beta4: actin-sequestering protein moonlights to repair injured tissues. *Trends Mol. Med.* **11:** 421–429.
4. Huff, T., C.S. Muller, A.M. Otto, *et al.* 2001. beta-Thymosins, small acidic peptides with multiple functions. *Int. J. Biochem. Cell Biol.* **33:** 205–220.
5. Sosne, G., P. Qiu & M. Kurpakus-Wheater. 2007. Thymosin beta 4: a novel corneal wound healing and anti-inflammatory agent. *Clin. Ophthalmol.* **1:** 201–207.
6. Sosne, G., P. Qiu, P.L. Christopherson & M.K. Wheater. 2007. Thymosin beta 4 suppression of corneal NFkappaB: a potential anti-inflammatory pathway. *Exp. Eye Res.* **84:** 663–669.
7. Sosne, G., P.L. Christopherson, R.P. Barrett & R. Fridman. 2005. Thymosin-beta4 modulates corneal matrix metalloproteinase levels and polymorphonuclear cell infiltration after alkali injury. *Invest Ophthalmol. Vis. Sci.* **46:** 2388–2395.
8. Young, J.D., A.J. Lawrence, A.G. MacLean, *et al.* 1999. Thymosin beta 4 sulfoxide is an anti-inflammatory agent generated by monocytes in the presence of glucocorticoids. *Nat. Med.* **5:** 1424–1427.

9. Girardi, M., M.A. Sherling, R.B. Filler, *et al.* 2003. Anti-inflammatory effects in the skin of thymosin-beta4 splice-variants. *Immunology* **109:** 1–7.

10. Badamchian, M., M.O. Fagarasan, R.L Danner, *et al.* 2003. Thymosin beta(4) reduces lethality and down-regulates inflammatory mediators in endotoxin-induced septic shock. *Int. Immunopharmacol.* **3:** 1225–1233.

11. Ben-Neriah, Y. & M. Karin. 2011. Inflammation meets cancer, with NF-kappaB as the matchmaker. *Nat. Immunol.* **12:** 715–723.

12. Ghosh, S. & M.S. Hayden. 2008. New regulators of NF-kappaB in inflammation. *Nat. Rev. Immunol.* **8:** 837–848.

13. Qiu, P., M.K. Wheater, Y. Qiu & G. Sosne. 2011. Thymosin beta4 inhibits TNF-alpha-induced NF-kappaB activation, IL-8 expression, and the sensitizing effects by its partners PINCH-1 and ILK. *FASEB J.* **25:** 1815–1826.

14. Sopko, N., Y. Qin, A. Finan, *et al.* 2011. Significance of thymosin beta4 and implication of PINCH-1-ILK-alpha-parvin (PIP) complex in human dilated cardiomyopathy. *PLoS One* **6:** e20184.

15. Ho, J.H., C.H. Chuang, C.Y. Ho, *et al.* 2007. Internalization is essential for the antiapoptotic effects of exogenous thymosin beta-4 on human corneal epithelial cells. *Invest. Ophthalmol. Vis. Sci.* **48:** 27–33.

16. Ho, J.H., K.C. Tseng, W.H. Ma, *et al.* 2008. Thymosin beta-4 upregulates anti-oxidative enzymes and protects human cornea epithelial cells against oxidative damage. *Br. J. Ophthalmol.* **92:** 992–997.

17. Kumar, S. & S. Gupta. 2011. Thymosin beta 4 prevents oxidative stress by targeting antioxidant and anti-apoptotic genes in cardiac fibroblasts. *PLoS One* **6:** e26912.

18. Bock-Marquette, I., A. Saxena, M.D. White, *et al.* 2004. Thymosin beta4 activates integrin-linked kinase and promotes cardiac cell migration, survival and cardiac repair. *Nature* **432:** 466–472.

19. Barnaeva, E., A. Nadezhda, E. Hannappel, *et al.* 2007. Thymosin beta4 upregulates the expression of hepatocyte growth factor and downregulates the expression of PDGF-beta receptor in human hepatic stellate cells. *Ann. N.Y. Acad. Sci.* **1112:** 154–160.

20. Cavasin, M.A., T.D. Liao, X.P. Yang, *et al.* 2007. Decreased endogenous levels of Ac-SDKP promote organ fibrosis. *Hypertension* **50:** 130–136.

21. Peng, H., O.A. Carretero, D.R. Brigstock, *et al.* 2003. Ac-SDKP reverses cardiac fibrosis in rats with renovascular hypertension. *Hypertension* **42:** 1164–1170.

22. Rhaleb, N.E., H. Peng, X.P. Yang, *et al.* 2001. Long-term effect of N-acetyl-seryl-aspartyl-lysyl-proline on left ventricular collagen deposition in rats with 2-kidney, 1-clip hypertension. *Circulation* **103:** 3136–3141.

23. De, S.M., R. Inzitari, S.L. Bosello, *et al.* 2011. beta-Thymosins and interstitial lung disease: study of a scleroderma cohort with a one-year follow-up. *Respir. Res.* **12:** 22.

24. Ashcroft, T., J.M. Simpson & V. Timbrell. 1988. Simple method of estimating severity of pulmonary fibrosis on a numerical scale. *J. Clin. Pathol.* **41:** 467–470.

Ann. N.Y. Acad. Sci. ISSN 0077-8923

Thymosin β4 affecting the cytoskeleton organization of the myofibroblasts

H. Paul Ehrlich and Sprague W. Hazard

Division of Plastic Surgery, Penn State University College of Medicine, Hershey, Pennsylvania

Address for correspondence: H. Paul Ehrlich, Ph.D., Division of Plastic Surgery, College of Medicine, The Pennsylvania State University, H071, 500 University Drive, Hershey, PA 17033. pehrlich@psu.edu

In previous studies, granulation tissue from subcutaneous sponge implants in rats receiving thymosin β4, a 43-amino acid actin-binding protein that advances wound repair, produced the unexpected absence of myofibroblast populations, along with uniform organized collagen fibers within the newly deposited connective tissue matrix. This result raised the question of whether the Tβ4 effect on blocking fibroblasts transformation into myofibroblasts a direct or indirect one. We report here work in progress to address this question. When human dermal fibroblasts are plated at low density, upon reaching confluence, they all express α smooth muscle actin (αSMA) within their cytoplasmic stress fibers, morphologically defining them as myofibroblasts. Treating low-density plated fibroblasts with Tβ4 prevents their expression of αSMA, as well as the generation an uneven distribution of microtubules within the cytoplasm. The speculation is that Tβ4 disruption of the distribution of microtubules alters the TGF-β–Smad signaling pathway, thus blocking fibroblast transformation into myofibroblasts.

Keywords: thymosin β4; myofibroblasts; stress fibers; microtubules

Introduction

Wound fibroblasts migrate into the site of tissue loss, where they become stationary, resident cells. These wound fibroblasts undergo a transition from the fibroblast phenotype into the myofibroblast phenotype. The myofibroblast, identified by the expression of α smooth muscle actin (SMA) within cytoskeletal stress fibers, is iconic of granulation tissue.[1] Myofibroblasts synthesize and deposit the connective tissue matrix of granulation tissue and generate the tension characteristic of granulation tissue through sustained contractile forces via their prominent cytoskeletal stress fibers.[2] In full-thickness wounds, myofibroblasts, through cell contractile generated forces, are alleged to produce the forces that cause wound closure via wound contraction. However, as previously reported vanadate- and thymosin beta 4 (Tβ4)-treated wounds do not contain myofibroblasts, yet wound contraction is unaffected by their absence.[3,4] Thus, the question arises: is the absence of αSMA-enriched stress fibers in myofibroblasts within granulation tissue a direct effect of Tβ4 or an indirect effect?

Tβ4, a 43-amino acid actin-binding protein, promotes wound repair.[5,6] Glucocorticoid-treated wounds show retarded wound healing, but treating such wounds with Tβ4 restores normal wound healing,[7] and it does so on many fronts. Possible mechanisms for Tβ4 enhancement of wound repair include down-regulation of inflammatory chemokines and cytokines, promotion of cell migration, increase in cell survival, and inhibition of microbial infiltration.[5,8] Tβ4 advances the migration of keratinocytes, enhances angiogenesis that promotes an enhanced vascular supply, and boosts the deposition of connective tissue by increasing collagen synthesis.[6,7]

Tβ4 forms a functional complex with integrin-linked kinase, which results in activation of the survival serine/threonine kinase (Akt), also known as protein kinase B (PKB).[9] Both Tβ4 and a naturally occurring peptide Ac-SDKP (amino acids 1–4 derived from Tβ4) prevent hypertension-induced

doi: 10.1111/j.1749-6632.2012.06730.x

cardiac fibrosis.[10,11] In part, the Ac-SDKP peptide is generated by prolyloligopeptidase, and inhibition of this enzyme is associated with an increased fibrotic response.[12,13] Although evidence indicates that Tβ4 reduces scarring from trauma to the heart, few reports reveal how Tβ4 affects scar formation in other tissues, such as skin.

The generation of scar tissue occurs through the maturation of granulation tissue during the remodeling phase of repair. To reduce the severity of scarring, the expectation is that this will occur by the deposition of a uniform connective tissue matrix in developing granulation tissue. The local injection of Tβ4 into subdermal polyvinyl alcohol (PVA) sponge implants in rats—an experimental model for the investigation of granulation tissue—has demonstrated that Tβ4 promotes uniform packing of collagen fibers within the newly deposited connective tissue matrix within granulation tissue. Tβ4-treated PVA sponge implants also have shown the replacement of myofibroblasts with fibroblasts. In granulation tissue from untreated PVA sponge implants, myofibroblasts, identified by expressing αSMA in cytoplasmic stress fibers, are the predominate cell type.[4] The organization of connective tissue (revealed by birefringence patterns from polarized light microscopy of Sirius red-stained tissues, as well as from transmission electron microscopy) in untreated control granulation tissue show collagen fibers arranged in random arrays. In contrast, PVA implants treated with Tβ4 showed collagen fibers arranged in uniform arrays.[4] To begin to address the question of whether inhibition of the transformation of fibroblasts into myofibroblasts by Tβ4 is a direct effect on fibroblasts or an indirect effect involving the activity of other cells, we are investigating Tβ4 in cell culture for its ability to transform myofibroblasts back into fibroblasts.

Upon reaching confluence, cultured human dermal fibroblasts initially plated at low density transform into myofibroblasts, which are identified by prominent cytoskeletal stress fibers containing αSMA.[14,15] TGF-β is proposed responsible for the changeover from the fibroblast phenotype to myofibroblast phenotype in these monolayer cultures.[14] In this study, myofibroblasts generated by low-density plated fibroblasts received Tβ4 to determine if the myofibroblasts revert back into fibroblasts in response to added Tβ4.

The experimental setup for our preliminary studies is the following. Fibroblasts plated on cover slips at low density and grown to confluence transform into myofibroblasts; half the myofibroblast cultures receive medium supplemented with 10 μg of Tβ4 (treated cells) and the other half receive phosphate buffered saline (control cells). After 24 h in culture, treated and control myofibroblasts are fixed, permeablized with detergent, and then stained by immunochemistry for αSMA to differentiate myofibroblasts from fibroblasts. Cells are also stained by immunochemistry for β-tubulin to identify microtubules; cell nuclei are stained fluorescent blue with DAPI, and filamentous-actin filaments are stained fluorescent green with Alexa phalloidin.[15] With a fluorescent microscope, we examine all cover slips and take photographs with a digital camera.

Our work in progress thus far shows that upon reaching confluence, low-density plated control fibroblasts had indeed transformed into myofibroblasts, which were identified by prominent fluorescent red αSMA stained cytoplasmic stress fibers. Figure 1A shows a confluent monolayer of myofibroblasts (control cells) with fluorescent blue nuclei and prominent fluorescent green cytoplasmic stress fibers enriched in fluorescent red αSMA—the signature of a myofibroblast.[1] Tβ4-treated myofibroblast cultures exhibited a different staining pattern (Fig. 1B), where cells had distinct fluorescent green cytoplasmic actin-rich stress fibers contained within spindle-shaped cells. However, unlike the control cells that were positive for αSMA, only a few of the Tβ4-treated cells demonstrated red fluorescence staining, signifying the loss of the myofibroblast phenotype.

We also noted changes between control and Tβ4-treated cells in the patterns of microtubules. Fluorescent microtubules in control myofibroblasts exhibited a uniform distribution throughout the cell cytoplasm (Fig. 2A). In contrast, the staining pattern of microtubules in myofibroblasts treated with Tβ4 exhibited a disrupted distribution of microtubules within their cytoplasm, consistent with their having reverted back to fibroblasts (Fig. 2B). In the Tβ4-treated cells, the major portion of microtubules was localized at the periphery of the nucleus, whereas few were localized at the periphery of the cells. This alteration in the pattern of the distribution of microtubules was a consistent finding in the Tβ4-treated cells.

A

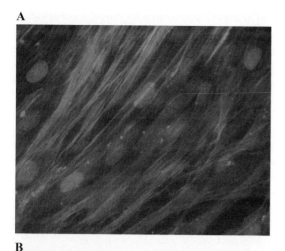

B

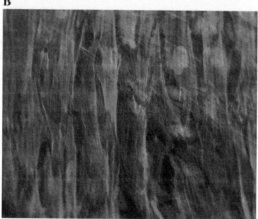

Figure 1. Low-density plated fibroblasts at confluence, stained for αSMA and filamentous-actin. To identify the myofibroblast phenotype in confluent low-density plated cells at 10 days, cells were immunostained for αSMA, demonstrated by red fluorescence, and filamentous-actin by phalloidin staining, demonstrated by green fluorescence. Cell nuclei were stained with DAPI, showing a blue fluorescence. Panel (A) shows nearly all the cells within the control culture are positive for αSMA, adjacent to thick actin-rich filaments, called stress fibers. Panel (B) shows low-density plated fibroblasts at confluence, myofibroblasts, which were treated for 24 h with 10 μg/mL of Tβ4 one day before processing for histological evaluations. Very few treated cells showed red fluorescence αSMA staining, supporting the notion that Tβ4 promoted the transformation of myofibroblasts back into fibroblasts.

What are the possible mechanisms for Tβ4 directing the deficiency in the expression of αSMA and the changeover from the myofibroblast phenotype to the fibroblast phenotype? At confluence, fibroblasts plated at moderate densities show only 15% of the cells expressing αSMA in stress fibers; however, when fibroblasts plated at moderate den-

sities receive TGF-β, a majority of the cells acquired the myofibroblast phenotype, expressing αSMA in stress fibers.[16] It has been reported that the mechanism for the transformation of fibroblasts plated at low density into myofibroblasts at confluence is through TGF-β.[14] This prompts the following question: What is the possible relationship among TGF-β, the disruption of the pattern of microtubules in myofibroblasts in monolayer culture treated with Tβ4, and the loss of αSMA expression within cytoskeletal stress fibers?

The demonstration of the fibrotic responses generated by TGF-β requires intracellular signaling by the Smad signaling pathway.[17] Although we were unable to find literature reporting a relationship between TGF-β and Tβ4, one possibility is Tβ4 disruption of the organization of microtubules may alter the Smad signaling pathway. It is well established that TGF-β promotes the transformation of fibroblasts into myofibroblasts.[16] TGF-β utilizes the Smad intracellular signaling pathway for initiating the transcription of specific genes, one of which includes the transcription of αSMA. Initially, TGF-β binds to a pair of specific receptors on the surface of fibroblasts. These receptors initiate intracellular responses, which results in the activation of the Smad intracellular signaling pathways. The initial step in the pathway involves the phosphorylation of Smad 3, which then forms a complex with Smad 2. The Smad 2/3 complex binds to Smad 4, generating the Smad 2/3–Smad 4 complex, which enters the nucleus and then activates specific genes, including αSMA.[17] Disruption of the Smad signaling pathway is expected to block the expression of αSMA.

The change in the pattern of the distribution of microtubules in myofibroblasts treated with Tβ4 suggests an involvement of TGF-β with gap junctional intercellular communications (GJIC). Both fibroblasts and myofibroblasts communicate directly with one another via GJIC.[18] Gap junctions between coupled cells require a specific structure on their surface called a *connexon*. A pair of connexons, each embedded within the surface of neighboring cells' plasma membranes, come together forming a channel or pore through which passes small molecules of less than 1000 MW directly between the cytosol of the coupled cells.[19] The combination of 6 connexin (Cx) proteins creates a connexon.

The family of Cx proteins includes Cx43, the major Cx present within fibroblasts.[20] The transport of

A

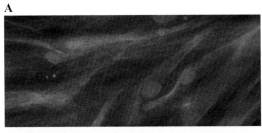

B

Figure 2. Low-density plated fibroblasts, myofibroblasts at confluence, stained for tubulin for the identification of microtubules. Microtubules were identified within cells by red fluorescence immunestaining of β-tubulin. Cell nuclei were stained fluorescent blue. Panel (A) shows microtubules within control myofibroblasts, which are uniformly distributed throughout the cell's cytoplasm. In contrast, as shown in panel (B), Tβ4-treated cells show an uneven distribution of microtubules. Most of the microtubules were near the cell nuclei, with an absence of microtubules at the periphery of the cells.

Cx43 from its site of synthesis at the endoplasmic reticulum to the cell's plasma membrane involves the direct attachment of Cx43 to microtubules at a specific binding site on the tubulin protein.[21] The specific binding site for Cx43 on tubulin is shared with Smad 3. The attachment of Cx43 to tubulin releases Smad 3 into the cytosol, where it is available to participate in the Smad signaling pathway.[21,22] Without the displacement of Smad 3 from tubulin by Cx43, Smad 2/3 remains sequestered on cytoplasmic microtubules, preventing its participation in the Smad signaling pathway. The disruption of microtubules at the cell's periphery will interfere with the transport of Cx43 and its release at the cell's membrane. The speculation is that retained Smad 3 on microtubules obstructs the Smad signaling pathway and prevents the transcription of αSMA and, further, that the direct effect of Tβ4 on the disruption of the distribution of microtubules is responsible for the loss of αSMA expression with the conversion of myofibroblasts back into fibroblasts.

An advantage of preventing the appearance of myofibroblasts within granulation tissue is the deposition of a more uniform organized connective tissue matrix within granulation tissue.[3] The absence of myofibroblasts in rat granulation tissue as reported in vanadate-treated wounds leads to the deposition of uniform organized collagen fibers by aligned fibroblasts.[3,23] The introduction of SB-505124, an agent that blocks the TGF-β/Smad signaling, eliminates the appearance of myofibroblasts in rat granulation, along with the deposition of a uniform organized connective tissue matrix of aligned collagen fibers.[24] Tβ4 also prevents the appearance of myofibroblasts in rat wound granulation tissue, with the outcome being the deposition of a uniform organized connective tissue matrix.[4] Interfering with the Smad signaling pathway eliminates the myofibroblast phenotype within granulation tissue and the deposition of a uniform organized connective tissue matrix. Thus, the mechanism for Tβ4 promotion of the myofibroblast phenotype reverting back into the fibroblast phenotype is likely through the disruption of the distribution of microtubules.

Conflicts of interest

The authors declare no conflicts of interest.

References

1. Gabbiani, G. 2003. The myofibroblast in wound healing and fibrocontractive diseases. *J. Pathol.* **200:** 500–503.
2. Hinz, B., D. Mastrangelo, C.E. Iselin, *et al.* 2001. Mechanical tension controls granulation tissue contractile activity and myofibroblast differentiation. *Am J Pathol.* **159:** 1009–1020.
3. Ehrlich, H.P., K.A. Keefer, G.O. Maish, 3rd, *et al.* 2001. Vanadate ingestion increases the gain in wound breaking strength and leads to better organized collagen fibers in rats during healing. *Plast Reconstr Surg.* **107:** 471–477.
4. Ehrlich, H.P., S.W. Hazard, 3rd. 2010. Thymosin beta4 enhances repair by organizing connective tissue and preventing the appearance of myofibroblasts. *Ann N Y Acad Sci.* **1194:** 118–124.
5. Goldstein, A.L., E. Hannappel & H.K. Kleinman. 2005. Thymosin beta4: actin-sequestering protein moonlights to repair injured tissues. *Trends Mol Med.* **11:** 421–429.
6. Malinda, K.M., G.S. Sidhu, H. Mani, *et al.* 1999. Thymosin beta4 accelerates wound healing. *J Invest Dermatol.* **113:** 364–368.
7. Philp, D., M. Badamchian, B. Scheremeta, *et al.* 2003. Thymosin beta 4 and a synthetic peptide containing its actin-binding domain promote dermal wound repair in db/db diabetic mice and in aged mice. *Wound Repair Regen.* **11:** 19–24.
8. Huang, L.C., D. Jean, R.J. Proske, *et al.* 2007. Ocular surface expression and in vitro activity of antimicrobial peptides. *Curr Eye Res.* **32:** 595–609.

9. Bock-Marquette, I., A. Saxena, M.D. White, *et al.* 2004. Thymosin beta4 activates integrin-linked kinase and promotes cardiac cell migration, survival and cardiac repair. *Nature.* **432:** 466–472.

10. Rossdeutsch, A., N. Smart & P.R. Riley. 2008. Thymosin beta4 and Ac-SDKP: tools to mend a broken heart. *J Mol Med.* **86:** 29–35.

11. Cavasin, M.A. 2006. Therapeutic potential of thymosin-beta4 and its derivative N-acetyl-seryl-aspartyl-lysyl-proline (Ac-SDKP) in cardiac healing after infarction. *Am J Cardiovasc Drugs.* **6:** 305–311.

12. Cavasin, M.A., T.D. Liao, X.P. Yang, J.J. Yang, *et al.* 2007. Decreased endogenous levels of Ac-SDKP promote organ fibrosis. *Hypertension.* **50:** 130–136.

13. Cavasin, M.A., N.E. Rhaleb, X.P. Yang, *et al.* 2004. Prolyl oligopeptidase is involved in release of the antifibrotic peptide Ac-SDKP. *Hypertension.* **43:** 1140–1145.

14. Masur, S.K., H.S. Dewal, T.T. Dinh, *et al.* 1996. Myofibroblasts differentiate from fibroblasts when plated at low density. *Proc Natl Acad Sci U S A.* **93:** 4219–4223.

15. Ehrlich, H.P., G.M. Allison & M. Leggett. 2006. The myofibroblast, cadherin, alpha smooth muscle actin and the collagen effect. *Cell Biochem Funct.* **24:** 63–70.

16. Desmouliere, A., A. Geinoz, F. Gabbiani, *et al.* 1993. Transforming growth factor-beta 1 induces alpha-smooth muscle actin expression in granulation tissue myofibroblasts and in quiescent and growing cultured fibroblasts. *J Cell Biol.* **122:** 103–111.

17. Derynck, R. & Y.E. Zhang. 2003. Smad-dependent and Smad-independent pathways in TGF-beta family signalling. *Nature.* **425:** 577–584.

18. Spanakis, S.G., S. Petridou & S.K. Masur. 1998. Functional gap junctions in corneal fibroblasts and myofibroblasts. *Invest Ophthalmol Vis Sci.* **39:** 1320–1328.

19. Goodenough, D.A. & D.L. Paul. 2009. Gap junctions. *Cold Spring Harb Perspect Biol.* **1:** a002576.

20. Bosco, D., J.A. Haefliger & P. Meda. 2011. Connexins: key mediators of endocrine function. *Physiol Rev.* **91:** 1393–1445.

21. Dai, P., T. Nakagami, H. Tanaka, *et al.* 2007. Cx43 mediates TGF-beta signaling through competitive Smads binding to microtubules. *Mol Biol Cell.* **18:** 2264–2273.

22. Giepmans, B.N., I. Verlaan, T. Hengeveld, *et al.* 2001. Gap junction protein connexin-43 interacts directly with microtubules. *Curr Biol.* **11:** 1364–1368.

23. Lee, M.Y. & H.P. Ehrlich. 2008. Influence of vanadate on migrating fibroblast orientation within a fibrin matrix. *J Cell Physiol.* **217:** 72–76.

24. Au, K. & H.P. Ehrlich. 2010. When the Smad signaling pathway is impaired, fibroblasts advance open wound contraction. *Exp Mol Pathol.* **89:** 236–240.

Ann. N.Y. Acad. Sci. ISSN 0077-8923

ANNALS OF THE NEW YORK ACADEMY OF SCIENCES
Issue: *Thymosins in Health and Disease*

Thymosin β4 stabilizes hypoxia-inducible factor-1α protein in an oxygen-independent manner

Mee Sun Ock,[1] Kyoung Seob Song,[2] Hynda Kleinman,[4] and Hee-Jae Cha[1,3]

[1]Department of Parasitology and Genetics, [2]Department of Physiology, [3]Institute for Medical Science, Kosin University College of Medicine, Busan, South Korea. [4]Craniofacial Developmental Biology and Regeneration Branch, National Institute of Dental and Craniofacial Research, National Institutes of Health, Bethesda, Maryland

Address for correspondence: Hee-Jae Cha, Department of Parasitology and Genetics, Kosin University College of Medicine, 34 Amnam-dong, Seo-gu, Busan 602-703, South Korea. hcha@kosin.ac.kr

The small actin-binding protein thymosin β4 (Tβ4) is understood to stimulate angiogenesis. Previously, we reported that Tβ4 induces angiogenesis by increasing vascular endothelial growth factor (VEGF) expression, but the mechanism underlying how Tβ4 upregulates VEGF expression remain unknown. To identify the mechanism of VEGF induction by Tβ4, we measured VEGF promoter activity and analyzed the effect of Tβ4 on VEGF RNA stability. The Tβ4 peptide had no effect on either VEGF promoter activity or VEGF RNA stability. We focused on the possibility that Tβ4 may indirectly induce VEGF expression via hypoxia-inducible factor (HIF)-1α. We determined that Tβ4 increased the stability of HIF-1α protein under normoxic conditions. These data suggest that Tβ4 indirectly induces VEGF expression by increasing the protein stability of HIF-1α in an oxygen-independent manner.

Keywords: thymosin β4; VEGF; HIF-1α; angiogenesis; normoxia; hypoxia

Thymosin β4 (Tβ4) plays multiple functional roles in cell physiology including accelerating cell migration,[1] angiogenesis,[1–3] wound healing,[4,5] hair growth,[5–7] tumor growth, and metastasis.[8,9] It also has been reported to suppress apoptosis[10–13] and inflammation.[14,15] Among these roles, angiogenesis is one of the most critical functions. Previous reports suggest various mechanisms for how Tβ4 stimulates angiogenesis. Some reports indicate that Tβ4 stimulates angiogenesis by inducing endothelial cell differentiation.[2] Other reports propose that Tβ4 stimulates the directional migration of human umbilical vein endothelial cells (HUVECs).[2,3] However, these findings are not sufficient to fully explain the story of Tβ4-mediated angiogenesis. We previously reported that Tβ4 induces the expression of vascular endothelial growth factor (VEGF), which is crucial for angiogenesis. However, this finding only partially explains Tβ4-mediated angiogenesis. Here we focus our current research on the mechanisms underlying the Tβ4 induction of VEGF expression.

Tβ4 stimulates angiogenesis

Angiogenesis involves a series of complex steps, including cell attachment, basement membrane degradation, cell migration, proliferation, and differentiation. Many compounds, including Tβ4, have been reported to induce angiogenesis.[16] Tβ4 stimulation of angiogenesis was initially identified via the rapid induction of Tβ4 genes after a culture of HUVECs on Matrigel™ (basement membrane matrix).[17] Tβ4 was also induced during endothelial cell differentiation and transfection of HUVECs with Tβ4 accelerated the rate of attachment spreading, and tube formation.[17] Another study demonstrated that Tβ4 stimulates the directional migration of HUVECs when used as a chemoattractant for endothelial cells.[3] In addition, Tβ4 significantly accelerated the rate of endothelial cell migration *in vivo* in a subcutaneous Matrigel plug assay.[3] Philp *et al.* reported that the actin-binding motif of Tβ4 was both necessary and sufficient for facilitating angiogenesis.[16] These studies provide insight on the

doi: 10.1111/j.1749-6632.2012.06657.x

idea that the angiogenic response involves Tβ4-mediated rearrangement of the actin cytoskeleton of endothelial cells. However, these findings do not fully explain the mechanism for angiogenic induction by Tβ4. Stimulation of endothelial cell migration by Tβ4 peptide after deletion of the LKKTET actin-binding motif was reduced relative to the intact peptide. However, even peptide without actin-binding motif still promoted endothelial cell migration comparing with not treated control. And only the actin-binding motif has less effect on stimulation of endothelial cell migration comparing with full length peptide. These results suggests that other regions of the peptide are also required for maximal effect. In addition, other β-thymosins, such as Tβ10, inhibit angiogenesis despite sharing the highly conserved actin-binding motif. Therefore, Tβ4-mediated angiogenesis resulting from the rearrangement of the actin cytoskeleton is likely only one part of the mechanism involved in angiogenic activity.

Tβ4 induces VEGF expression

While studying the effect of Tβ4 on tumor growth and metastasis, we determined that tumors from Tβ4 overexpressing B16-F10 melanoma contained more blood vessels relative to the controls.[9] To identify the effect of Tβ4 on the expression of angiogenesis-related factors, we compared the expressions of angiogenic growth factors with Tβ4 overexpressing adenovirus-infected B16-F10 melanoma cells as well as controls. Using the Western blot analysis of angiogenesis-related proteins, we determined that the expression level of VEGF was significantly upregulated, but other factors, including fibroblast growth factors (FGFs), epidermal growth factor (EGF), platelet-derived growth factor (PDGF), and insulin-like growth factor (IGF), were not upregulated. Furthermore, Tβ4 peptide treatment also induced VEGF expression[9] but did not affect the expression of other growth factors. These findings suggest that Tβ4 increased VEGF levels, in turn stimulating angiogenesis. Another report demonstrated that Tβ10 inhibited angiogenesis by reducing the VEGF expression levels, and it suggests that the induction of VEGF expression by Tβ4 plays one of the key roles in angiogensis.[18] It also explains how both Tβ4 and Tβ10 have the same actin-binding motif but show different functions. Therefore, Tβ4 stimulates additionally endothelial

migration but also induces VEGF expression to stimulate angiogenesis.

Molecular mechanisms of VEGF expression via Tβ4

How Tβ4 induces VEGF expression remains an unsolved question. There are many possible hypotheses. Tβ4 can bind to a receptor and stimulate the signal transduction cascades or go directly into the cytoplasm and/or nucleus and stimulate further pathways involved in VEGF expression. Previous reports indicate that Tβ4 can enter the nucleus directly[19] and modulate VEGF expression. These hypotheses are based on the fact that Tβ4 stimulates the promoter activity of VEGF. Another possibility for VEGF induction by Tβ4 involves increased mRNA stability by Tβ4 without affecting VEGF promoter activity. To clarify the mechanisms of VEGF induction by Tβ4, we used a luciferase assay system with VEGF in the 5' promoter region. Using this approach, we determined that Tβ4 peptide had no effect on VEGF promoter activity.[20] We also assessed the effect of Tβ4 on the mRNA stability of VEGF, but found no effect on VEGF mRNA stability.[20] To explain these unexpected findings, we analyzed the possibility that Tβ4 induces VEGF indirectly through HIF-1α, one of the most critical transcription factors in upregulating VEGF expression (Fig. 1). Cotreatment of HIF-1 and Tβ4 expression vectors demonstrated a significant increase in VEGF promoter activity compared with HIF-1 from untreated cells.[20] Treatment of the Tβ4 peptide with HIF-1 expression vectors also increased VEGF

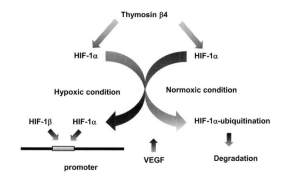

Figure 1. Proposed mechanism of Tβ4-mediated VEGF induction. Tβ4 stimulates the induction, stability, or activity of HIF-1α. HIF-1α binds to HIF-1β on the hypoxia response element (HRE) in the VEGF promoter region, triggering the expression of VEGF mRNA.

promoter activity significantly as compared with HIF-1 overexpressing vectors from untreated cells.[20] These results suggest that Tβ4 induced VEGF expression via HIF-1α. To clarify whether Tβ4 affects the HIF-1α expression at the mRNA or protein level, we conducted real-time polymerase chain reaction (PCR) and Western blot assays. Tβ4 increased the protein expression of HIF-1α, but it exhibited no effect at the mRNA level.[20] The HIF-1α protein was upregulated without induction at the mRNA level, suggesting that Tβ4 may increase the protein stability of the HIF-1α protein. An additional experiment with cyclohexamide that blocked the translation and addition of Tβ4 as a way to verify the protein stability of HIF-1α demonstrated that the HIF-1α level increased in the presence of Tβ4 and cyclohexamide,[20] but rapidly decreased when treated with just cyclohexamide.[20] These data suggested that Tβ4 increases HIF-1α protein stability and induces VEGF expression. We conclude that Tβ4 increases VEGF expression in an indirect manner by increasing HIF-1α protein stability, which sets the stage for increased VEGF induction by Tβ4.

Tβ4 stimulates HIF-1α stability in an oxygen-independent manner

The regulation of HIF-1α stability is a complex process. Both HIF-1α and HIF-2α are transcription factors that trigger the transactivation of target genes. They have two transcriptional activation domains, the N-terminal transactivation domain (NTAD) and the C-terminal transactivation domain (CTAD),[21] that enable transcription. During oxygen-dependent activity, HIF-1α activity and stability are controlled differently on the basis of normoxic and hypoxic conditions. Under normoxic conditions, the ability of HIF-α to activate transcription is prevented through HIF hydroxylation, which is catalyzed by prolyl hydroxylases (PHDs) near the NTAD and factor-inhibiting HIF (FIH) at the CTAD (Fig. 2). The von Hippel–Lindau (VHL) tumor suppressor protein (pVHL) binds to hydroxylated HIF-α and, with a complex containing elongin-C, elongin-B, cullin-2, and ring-box 1 (RBX1) of an E3 ubiquitin ligase, targets HIF-α for ubiquitination and degradation by the 26S proteasome.[21] Under hypoxic conditions, HIF-α hydroxylation by PHDs or FIH is prevented. However, FIH may remain active when compared with PHDs at lower oxygen concentrations, thereby suppressing the activity of

HIF-α proteins that avoid degradation under moderate hypoxia.[21] During hypoxia, unhydroxylated HIF-α cannot bind pVHL and therefore accumulates in the cell.[21] HIF-1α and HIF-2α translocate to the nucleus and heterodimerize with HIF-β. The HIF-α/HIF-β complex binds to hypoxia-response elements (HREs) on nuclear DNA and promotes the transcription of target genes involved in cell growth, angiogenesis, glucose metabolism, pH regulation, and cell survival/apoptosis.[21]

During oxygen-independent activity, VHL-independent HIF degradation occurs. HIF-1α degradation is also regulated in an oxygen/PHD/pVHL-independent manner by heat shock protein 90 (HSP90), which protects proteins from misfolding and degradation through its ATPase activity. Receptor for activated C-kinase 1 (RACK1), a novel HIF-1α–interacting protein, can promote proteasome-dependent degradation of HIF-1α in an oxygen/PHD/pVHL-independent manner and competes with HSP90 to bind to the PAS-A domain of HIF-1α (Fig. 2). HSP90 binding stabilizes HIF-1α by excluding RACK1, which binds elongin-C and recruits elongin-B and other components of E3 ubiquitin ligase to HIF-1α.[21]

Here, we have focused on the mechanism underlying Tβ4-mediated HIF-1α stabilization relative to oxygen-dependent and oxygen-independent activity. As indicated in a previous report, Tβ4 expression was not induced in B16-F10 cells under hypoxic conditions. However, the expression level of VEGF and HIF-1α were significantly induced under hypoxic conditions. Furthermore, increased HIF-1α stability as a result of Tβ4 interaction was evident under normoxic conditions.[20] These results strengthen the possibility that Tβ4 increases HIF-1α protein stability in an oxygen- independent manner. However, further studies are need to fully understand the influence of Tβ4 on HIF-1α protein stability.

Further studies

The main pathway of Tβ4-mediated angiogenesis was reported to be accomplished by induction of VEGF expression. In addition, Tβ4 also stimulates endothelial cell migration and differentiation by interaction with G-actin. However, the possibility of other pathways for Tβ4-mediated angiogenesis still exists. Tβ4 has been reported to induce other genes except VEGF, including laminin-5,[22] plasminogen activator inhibitor-1 (PAI-1),[23]

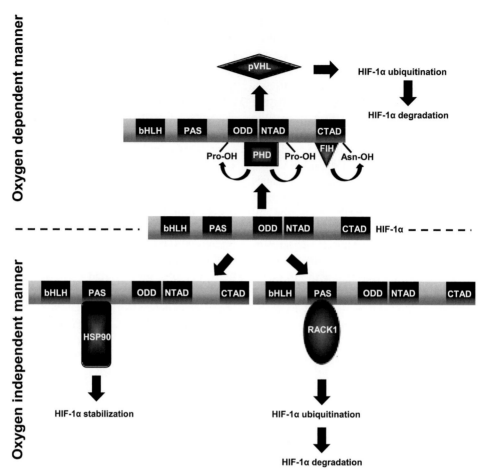

Figure 2. Oxygen-dependent and oxygen-independent regulation of HIF-1α. HIF-1α contains a basic helix–loop–helix PAS DNA-binding domain (bHLH) and two transactivation domains (NTAD and CTAD). In the presence of oxygen, PHDs hydroxylate proline residues (Pro) within the HIF-1α oxygen-dependent degradation domain (ODD) near the NTAD. pVHL binds to hydroxylated HIF-1α and elongin-C. Consequently, HIF-α is ubiquitinated and degraded by the 26S proteasome. Oxygen also influences the transcriptional activity of HIF-1α/HIF-2α through factor-inhibiting HIF (FIH), which hydroxylates an asparaginyl residue (ASN) within the HIF-α CTAD. During hypoxia, both HIF-1α and HIF-2α are stabilized and enter the nucleus, where they heterodimerize with HIF-β and initiate downstream transcription. In the oxygen-independent pathway, heat shock protein 90 (HSP90) and the receptor for activated C-kinase 1 (RACK1) compete to bind to the PAS domain of HIF-1α. HSP90 binding stabilizes HIF-1α by excluding RACK1, which recruits elongin-C to HIF-1α. This interaction promotes the polyubiquitination and degradation of HIF-1α, whereas HIF-2α can still activate downstream transcription. Thus, both oxygen-dependent and oxygen-independent pathways bring HIF-1α to elongin-C and other subunits of the E3 ubiquitin ligase complex for degradation by the 26S proteasome.

Zyxin,[24] and MMPs.[25] Tβ4 also goes into the cell nucleus and may alter gene regulation,[19] indicating that Tβ4 may induce angiogenesis through another pathways. To reveal the full pathways of Tβ4-mediated angiogenesis, gene or protein profiles should be analyzed by cDNA microarray or proteomics analysis.

Direct evidence that HIF-1α mediates induction of VEGF by Tβ4 should be confirmed by showing that inhibition of HIF-1α blocks the induction of VEGF by Tβ4. In addition, detection of Tβ4-mediated HIF-1α stability was conducted under only normoxic condition; interaction of Tβ4 and HIF-1α in hypoxic condition should also be investigated. Tβ4-induced angiogenesis or VEGF gene expression should be also analyzed under hypoxic condition. Furthermore, detailed studies showing how Tβ4 increases HIF-1α stability should be performed under normoxic and hypoxic conditions, including interaction of Tβ4 with key regulatory factors of HIF-1α stability such as VHL, Hsp90, and RACK1 under both conditions.

Acknowledgment

This work was supported by the Korea Research Foundation Grant KRF-20110003918, funded by the Korean Government.

Conflicts of interest

The authors declare no conflicts of interest.

References

1. Van Troys, M., D. Dewitte, M. Goethals, *et al.* 1996. The actin binding site of thymosin beta 4 mapped by mutational analysis. *EMBO J.* **15:** 201–210.
2. Grant, D.S., W. Rose, C. Yaen, *et al.* 1999. Thymosin beta4 enhances endothelial cell differentiation and angiogenesis. *Angiogenesis* **3:** 125–135.
3. Malinda, K.M., A.L. Goldstein & H.K. Kleinman. 1997. Thymosin beta 4 stimulates directional migration of human umbilical vein endothelial cells. *FASEB J.* **11:** 474–481.
4. Malinda, K.M., G.S. Sidhu, H. Mani, *et al.* 1999. Thymosin beta4 accelerates wound healing. *J. Invest. Dermatol.* **113:** 364–368.
5. Philp, D., M. Nguyen, B. Scheremeta, *et al.* 2004. Thymosin beta4 increases hair growth by activation of hair follicle stem cells. *FASEB J.* **18:** 385–387.
6. Philp, D., A.L. Goldstein & H.K. Kleinman. 2004. Thymosin beta4 promotes angiogenesis, wound healing, and hair follicle development. *Mech. Ageing Dev.* **125:** 113–115.
7. Philp, D., S. St-Surin, H.J. Cha, *et al.* 2007. Thymosin beta 4 induces hair growth via stem cell migration and differentiation. *Ann. N. Y. Acad. Sci.* **1112:** 95–103.
8. Kobayashi, T., F. Okada, N. Fujii, *et al.* 2002. Thymosin-beta4 regulates motility and metastasis of malignant mouse fibrosarcoma cells. *Am. J. Pathol.* **160:** 869–882.
9. Cha, H.J., M.J. Jeong & H.K. Kleinman. 2003. Role of thymosin beta4 in tumor metastasis and angiogenesis. *J. Natl. Cancer Inst.* **95:** 1674–1680.
10. Hall, A.K. 1994. Molecular interactions between G-actin, DNase I and the beta-thymosins in apoptosis: a hypothesis. *Med. Hypotheses* **43:** 125–131.
11. Iguchi, K., Y. Usami, K. Hirano, *et al.* 1999. Decreased thymosin beta4 in apoptosis induced by a variety of antitumor drugs. *Biochem. Pharmacol.* **57:** 1105–1111.
12. Niu, M. & V.T. Nachmias. 2000. Increased resistance to apop-

tosis in cells overexpressing thymosin beta four: A role for focal adhesion kinase pp125FAK. *Cell Adhes. Commun.* **7:** 311–320.
13. Moon, E.Y., J.H. Song & K.H. Yang. 2007. Actin-sequestering protein, thymosin-beta-4 (TB4), inhibits caspase-3 activation in paclitaxel-induced tumor cell death. *Oncol. Res.* **16:** 507–516.
14. Young, J.D., A.J. Lawrence, A.G. MacLean, *et al.* 1999. Thymosin beta 4 sulfoxide is an anti-inflammatory agent generated by monocytes in the presence of glucocorticoids. *Nat. Med.* **5:** 1424–1427.
15. Sosne, G., P. Qiu, P.L. Christopherson, *et al.* 2007. Thymosin beta 4 suppression of corneal NFkappaB: a potential anti-inflammatory pathway. *Exp. Eye Res.* **84:** 663–669.
16. Philp, D., T. Huff, Y.S. Gho, *et al.* 2003. The actin binding site on thymosin beta4 promotes angiogenesis. *FASEB J.* **17:** 2103–2105.
17. Grant, D.S., J.L. Kinsella, M.C. Kibbey, *et al.* 1995. Matrigel induces thymosin beta 4 gene in differentiating endothelial cells. *J. Cell Sci.* **108**(Pt 12): 3685–3694.
18. Zhang, T., X. Li, W. Yu, *et al.* 2009. Overexpression of thymosin beta-10 inhibits VEGF mRNA expression, autocrine VEGF protein production, and tube formation in hypoxia-induced monkey choroid-retinal endothelial cells. *Ophthalmic Res.* **41:** 36–43.
19. Huff, T., O. Rosorius, A.M. Otto, *et al.* 2004. Nuclear localisation of the G-actin sequestering peptide thymosin beta4. *J. Cell Sci.* **117:** 5333–5341.
20. Jo, J.O., S.R. Kim, M.K. Bae, *et al.* 2010. Thymosin beta4 induces the expression of vascular endothelial growth factor (VEGF) in a hypoxia-inducible factor (HIF)-1alpha-dependent manner. *Biochim. Biophys. Acta* **1803:** 1244–1251.
21. Baldewijns, M.M., I.J. van Vlodrop, *et al.* 2010. VHL and HIF signalling in renal cell carcinogenesis. *J. Pathol.* **221:** 125–138.
22. Sosne, G., L. Xu, L. Prach, *et al.* 2004. Thymosin beta 4 stimulates laminin-5 production independent of TGF-beta. *Exp. Cell Res.* **293:** 175–183.
23. Al-Nedawi, K.N., M. Czyz, R. Bednarek, *et al.* 2004. Thymosin beta 4 induces the synthesis of plasminogen activator inhibitor 1 in cultured endothelial cells and increases its extracellular expression. *Blood* **103:** 1319–1324.
24. Moon, H.S., S. Even-Ram, H.K. Kleinman, *et al.* 2006. Zyxin is upregulated in the nucleus by thymosin beta4 in SiHa cells. *Exp. Cell Res.* **312:** 3425–3431.
25. Philp, D., B. Scheremeta, K. Sibliss, *et al.* 2006. Thymosin beta4 promotes matrix metalloproteinase expression during wound repair. *J. Cell Physiol.* **208:** 195–200.

Ann. N.Y. Acad. Sci. ISSN 0077-8923

ANNALS OF THE NEW YORK ACADEMY OF SCIENCES
Issue: *Thymosins in Health and Disease*

Thymosin β4 and cardiac protection: implication in inflammation and fibrosis

Sudhiranjan Gupta,[1] Sandeep Kumar,[2] Nikolai Sopko,[3] Yilu Qin,[4] Chuanyu Wei,[1] and Il-Kwon Kim[1]

[1]Texas A & M Health Science Center, College of Medicine, Department of Medicine, Cardiovascular Research Institute; Scott and White, Central Texas Veterans Health Care System, Temple, Texas. [2]Department of Biomedical Engineering and Cardiology, Emory University, Atlanta, Georgia. [3]Case Western Reserve University, Cleveland, Ohio. [4]Cleveland Clinic, Lerner College of Medicine, Cleveland, Ohio

Address for correspondence: Dr. Sudhiranjan Gupta, Scott and White, Central Texas Veterans Health Care System, 1901 S. 1st Street, Building 205, Temple, TX 76504. sgupta@medicine.tamhsc.edu

Thymosin beta 4 (Tβ4) is a ubiquitous protein with diverse biological functions. The effecter molecules targeted by Tβ4 in cardiac protection remain unknown. We summarize previously published work showing that treatment with Tβ4 in the myocardial infarction setting improves cardiac function by activating Akt phosphorylation, promoting the ILK–Pinch–Parvin complex, and suppressing NF-κB and collagen synthesis. In the presence of Wortmannin, Tβ4 showed minimal cardiac protection. *In vitro* findings revealed that pretreatment with Tβ4 resulted in reduction of intracellular ROS in the cardiac fibroblasts and was associated with increased expression of antioxidant enzymes, reduction of Bax/Bcl$_2$ ratio, and attenuation of profibrotic genes. Silencing of Cu/Zn-SOD, catalase, and Bcl$_2$ genes abrogated the protective effect of Tβ4. Our findings suggest that Tβ4 improves cardiac function by enhancing Akt and ILK activation and suppressing NF-κB activity and collagen synthesis. Furthermore, Tβ4 selectively upregulates catalase, Cu/Zn-SOD, and Bcl$_2$, thereby protecting cardiac fibroblasts from H$_2$O$_2$-induced oxidative damage. Further studies are warranted to elucidate the signaling pathway(s) involved in the cardiac protection afforded by Tβ4.

Keywords: thymosin β4; NF-κB; cardiac fibrosis; antioxidative genes

Introduction

Thymosin beta 4 (Tβ4), originally isolated and characterized from the thymus gland, contributes important roles to G-actin sequestration, epithelialization, angiogenesis, stimulation of adult epicardial stem cell differentiation, prevention of cell death, and anti-inflammation.[1–5] Tβ4 is a 43 amino-acid small peptide that primarily interacts with G-actin and functions as an actin-sequestering protein in a majority of cell types.[6–8] Tβ4 is found in high concentrations in platelets, white blood cells, wound fluids, and in other tissues of the body. Tβ4 lacks a secretion signal, and it is assumed that the presence of Tβ4 in body fluids or in the site of damage is mainly due to injured tissues. Although Tβ4 may be released to the site of injury, it is not considered a growth factor. Tβ4 does not promote cell growth or bind to heparin, and it is ubiquitously present in most tissues.[3,9]

The past decade has witnessed the identification of a wide range of activities by Tβ4 that include wound repair, angiogenesis, cell migration, the survival mechanism, stem cell differentiation, and, importantly, cardiac repair. Incisional wounds, for example, can be repaired by Tβ4 in a rat model by rearranging connective tissues and lowering myofibroblasts formation.[10] Further, the fact that alkali-induced corneal wounds can be prevented by the treatment of Tβ4 seems to be due to the reduction of inflammation and apoptosis and the acceleration of corneal reepithelialization.[11–13] Interestingly, in recent years Tβ4 has been shown to promote stem cell differentiation involved in repairing the injured myocardium.[14–17] Tβ4 can induce endothelial progenitor cell (EPC) directional migration that is

doi: 10.1111/j.1749-6632.2012.06752.x

Ann. N.Y. Acad. Sci. 1269 (2012) 84–91 © 2012 New York Academy of Sciences.

essential for EPC-mediated reendothelialization and neovascularization in an ischemia/reperfusion model.[18–20] These studies indicate that EPCs are capable of providing cardiac protection in a Tβ4-dependent fashion.

It remains unknown how Tβ4 elicits diverse biological functions in different experimental settings; it is postulated that Tβ4 is taken up by the cells, internalized, and then acts intracellularly to promote various biological activities.

Role of Tβ4 in cardiac remodeling

Cardiac remodeling is an important pathological feature of structural alteration that contributes to ventricular stiffness, diastolic dysfunction, and arrhythmia.[21,22] Recently, Tβ4 has been investigated as a therapeutic agent for the treatment of ischemic heart disease in the hope that it may have significant therapeutic potential to protect the myocardium and to promote cardiomyocyte survival in the acute stages of ischemic heart disease.[14,15,23–25] Interestingly, in cardiac coronary intervention, Tβ4 level is increased after percutaneous coronary intervention in patients with ST-elevation acute myocardial infarction (MI), indicating a new role for Tβ4 in the assessment of reperfusion success after coronary angiography.[26] To investigate the underlying molecular mechanism of cardiac protection afforded by Tβ4, Bock-Marquette *et al.* identified PINCH-1 and activation of Akt and integrin link kinase (ILK) as possible mediators in this process.[14,23] In light of the role of ILK in cardiac pathologies, recent studies have shown that transgenic mice with cardiac-specific overexpression of ILK exhibited compensated cardiac hypertrophy,[27] whereas cardiac specific ablation of ILK revealed a dilated cardiomyopathy phenotype that advanced to cardiac failure, indicating a pivotal role of ILK in cardiac pathologies.[28] Subsequent analyses have further established a cardioprotective role of ILK in murine models of heart failure.[29–32]

ILK is a multidomain integrin adaptor protein that possesses widely conserved structural and signal transduction functions.[33] ILK forms a tripartite complex with the LIM domain protein PINCH-1 and with α-parvin.[34,35] Assembly of this complex is required for localization of ILK to focal and fibrillar adhesions, as well as for cell shape, motility, and survival.[35]

Table 1. The effect of Tβ4 on post-MI mice and quantification of the PIP complex

	WT	MI	MI + Tβ4
ILK	38.7 ± 2.48	148.24 ± 1.87	166.71 ± 1.41
Pinch-1	31.2 ± 1.03	108.73 ± 2.51	135.24 ± 1.73
α-Parvin	29.58 ± 2.54	107.94 ± 3.91	117.54 ± 2.6
pAkt	21.75 ± 0.69	65.02 ± 1.48	105.27 ± 1.49

We investigated the status of the PINCH-1–ILK–α-parvin (PIP) complex in human heart failure subjects.[36] We found that the PIP complex was significantly elevated along with Akt phosphorylation in failing human hearts, characterized as dilated cardiomyopathy.[36] To corroborate the data in an experimentally induced heart failure model such as MI, we observed significant enhancement of the PIP complex (Table 1). In the setting of Tβ4 treatment in post-MI, we observed moderate enhancement of the PIP complex and significant increment of Akt phosphorylation, which may be associated with improved cardiac function (Table 1). The induction of the PIP complex at this point may be evaluated as a consequence of stress- or injury-related phenomena, where it acts as a signal to regulate the remodeling process. To ascertain the mechanism of Tβ4 in cardiac repair, we evaluated whether Akt is critical in cardiac protection. We injected Wortmannin (an Akt inhibitor) in the presence or absence of Tβ4. We observed that treatment with Wortmannin in addition to Tβ4 abrogated the infarct-attenuating effects of Tβ4, resulting reappearance of infarction, comparable to MI (Fig. 1).[36] However, Wortmannin treatment minimally attenuated the protective effects of Tβ4 on cardiac function after MI. The untreated MI mice showed decreased cardiac function and increased left ventricular (LV) mass compared with sham-operated mice. Mice receiving Tβ4 showed significantly improved cardiac function compared with the untreated MI group. Treatment with Wortmannin moderately attenuated the beneficial effect of Tβ4 in an MI model (Table 2). This suggests that the beneficial effects of cardiac protection by Tβ4 in the post-MI period may not solely be dependent on increased Akt activation. Increased Akt activation may be, in part, responsible for the decreased

A

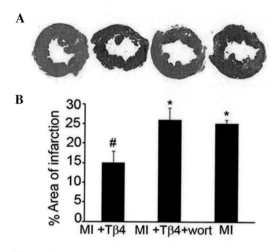

B

Figure 1. Infarct size analysis of Tβ4 and Tβ4 + Wortmannin–treated MI mice. (A) The percent infarct size was determined by Masson's trichome staining. A representative image of hearts after MI in the presence and absence of Tβ4 and Wortmannin. (B) Quantification of infarct areas. The error bars represent the average ± SE of five different mice in an independent experiment (# *P* = 0.02 and was significant in the comparison of MI vs. MI + Tβ4; * *P* = 0.69 and considered as nonsignificant in a one-way ANOVA analysis in the comparison of MI versus MI + Tβ4 + Wortmannin). Data in this figure are from Ref. 36.

infarct size, which is directly related to myocyte survival.

The significance and contribution of Tβ4 in cardiac development as well as repair after injury are well documented.[14,15] Tβ4 is shown to provide cardiac protection via mobilization of endothelial cells or progenitor cells after reperfusion injury in a mouse model.[20] In related studies, another group showed that Tβ4 can activate epicardial progenitor cells after MI,[37] and Tβ4 induced epicardial progenitor cell mobilization and neovascularization.[15] However, recent reports contradict these findings. Zhou *et al.* reported that Tβ4 treatment after MI did not reprogram epicardial cells into

cardiomyocytes.[38] The reason for the discrepancy may be due to the different timing of Tβ4 injection. Yet, the different experimental results raise caution and should drive further exploration of the potential beneficial effects of Tβ4 in cardiac protection. Another study, using a global knockout of Tβ4, showed that Tβ4 is indispensable in cardiac development.[39,40] Interestingly, the different results may be due to the technologies used that reveal the phenotype; one group used sh-RNA–mediated knockdown of Tβ4,[15] whereas the other group used a complete knockdown of Tβ4.[39]

Role of Tβ4 in inflammation

Inflammation contributes a key role in various pathologies of cardiac remodeling, including MI or ischemia/reperfusion injury, hypertrophy, and myocarditis.[41] Recent evidence indicates that several proinflammatory responses triggered by tissue injury or damage are associated with a multiprotein complex called the inflammasome.[42] Inflammatory mediators primarily include the interleukins and the tumor necrosis factor family membercytokines, which play a significant role in adverse cardiac remodeling.[42] Tβ4 has been shown to reduce inflammatory responses in various cells and in tissue injury models.[43–45] In a rat corneal epithelial debridement model, Tβ4 attenuated proinflammatory mediators and chemokines.[11–13,46,47] NF-κB, another key regulator of inflammation and oxidative stress, directly modulates various cardiac diseases, as mentioned earlier.[48,49] Tβ4 is effective in reducing NF-κB activation in TNF-α–stimulated corneal epithelial cells, suggesting clinical potential of Tβ4 as an anti-inflammatory agent.[50,51]

Recently, we investigated whether Tβ4 has any influence in attenuating inflammatory responses in a mouse MI model.[36] We observed that Tβ4 significantly reduced NF-κB activity in post-MI mice,

Table 2. The echocardiogram analysis of sham, MI, Tβ4-treated MI, and MI + Tβ4 + Wortmannin–treated mice after seven days of Tβ4 treatment

	WT	MI	MI +Tβ4	MI + Tβ4 + Wortmannin
LVM (mg)	87.7 ± 8.9	178.5 ± 12.2	109.6 ± 5.1	123.6 ± 12.9
EDD (mm)	3.18 ± 0.24	5.67 ± 0.39	3.97 ± 0.33	4.1 ± 0.33
EF (%)	87.0 ± 0.03	0.14 ± 0.06	0.52 ± 0.08	0.51 ± 0.02
FSH (%)	0.50 ± 0.05	0.10 ± 0.05	0.27 ± 0.06	0.21 ± 0.01

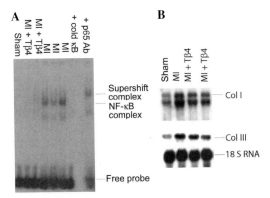

A

Supershift
complex
NF-κB
complex

Free probe

B

Col I

Col III

18 S RNA

Figure 2. Tβ4 inhibits NF-κB activation and attenuates cardiac fibrosis in MI mice. (A) Mice were subjected to MI; Tβ4 was injected and mice were kept for seven days. NF-κB activation was measured by gel mobility shift assay using ^{32}P NF-κB DNA as a probe. Cold NF-κB DNA was used for competition analysis. A p65 antibody was used for supershift analysis. (B) Collagen type I and type III mRNA expression was performed using their respective cDNA probes. Results are presented as five different mice ($n = 5$, $^{*}P < 0.001$ compared to sham or MI groups, Student's t-test). The one-way ANOVA test among the groups of MI and MI + Tβ4 also showed significant at $P < 0.05$ level. Data in this figure are from Ref. 36.

compared with untreated MI mice (Fig. 2A), indicating that cardiac protection was partly due the inhibition of the inflammatory response regulated by NF-κB.[36] Inhibition of NF-κB by Tβ4 may additionally support cardiac protection by

reducing the inflammatory response. This was the first report showing that Tβ4 suppresses NF-κB activation in an experimentally induced mouse MI model.[36]

Role of Tβ4 in fibrosis

Cardiac fibrosis is an important pathological feature of structural remodeling that contributes to ventricular stiffness, diastolic dysfunction, and arrhythmia.[52] The molecular and cellular basis for fibrosis is an uncontrolled accumulation of collagens along with other extracellular matrix (ECM) components, including matrix metalloproteinase (MMP), fibronectin, and connective tissue growth factor (CTGF), in the interstitial and perivascular regions of the heart.[21,53,54] As a result, myocardial stiffness occurs, which alters the mechanics and impairs the function of the heart. Accumulating evidence points to the antifibrotic properties of acetylated SDKP (Ac-SDKP) when it is administered chronically in rat models of hypertension and MI.[55,56] It has been reported that Tβ4 is the most likely precursor of this tetrapeptide (Ac-SDKP) because it possesses the sequence Ac-SDKP in its N-terminus, which would be sufficient to release the tetrapeptide.[44] However, the role Tβ4 in the context of cardiac fibrosis remains elusive. Recently, we found that Tβ4 has the ability to attenuate profibrotic gene

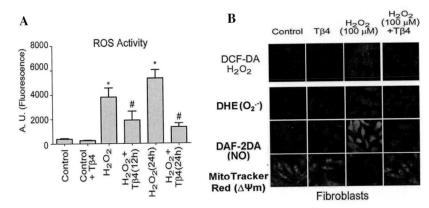

Figure 3. Effect of Tβ4 on generation of ROS in fibroblasts treated with H_2O_2 by fluorimetry. (A) The graph represents the percentage of fluorescence-positive fibroblasts upon staining with DCF-DA. Data represent the mean ± SE of at least three separate experiments. Note that * indicates $P < 0.05$ compared to the controls and # indicates $P < 0.05$ compared to the respective H_2O_2-treated group. (B) First and second panel show the representative confocal laser scanning microscopy images of cells stained with DHE red, showing the effect of Tβ4 on generation of superoxide radicals upon treatment with H_2O_2 in fibroblasts. The third panel shows the representative confocal laser scanning microscopy images of cells stained with DAF-2DA showing the effect of Tβ4 on generation of nitric oxide upon treatment with H_2O_2 in fibroblasts. The last panel displays the representative confocal laser scanning microscopy images of cells stained with MitotrackerRed™, showing the effect of Tβ4 on loss of mitochondrial membrane potential upon treatment with H_2O_2 in fibroblast. Data in this figure are from Ref. 63.

expression in a post-MI mouse model.[63] Tβ4 significantly reduced the expression of collagen type I and type III—the hallmark for cardiac fibrosis—compared with untreated MI mice (Fig. 2B). This suggests that Tβ4 mitigated cardiac remodeling by reducing collagen deposition.

Role of Tβ4 against oxidative stress in cardiac fibroblasts

Increased oxidative stress is involved in the pathology of diverse diseases, including cardiac remodeling. Oxidative stress is defined as an imbalance in antioxidant defense mechanisms that elicit the production of reactive oxygen species (ROS).[57,58] ROS are primarily characterized as oxygen-based free chemical particles, and, if present in excess, they can cause contractile dysfunction and structural damage to the myocardium.[59] Moreover, it has been reported that increased levels of oxidative stress in the failing

heart are primarily due to the functional uncoupling of the respiratory chain caused by inactivation of complex I in mitochondria and considered to be a good source for ROS production.[60,61] With regard to the role of Tβ4 in oxidative stress, Ho *et al.* reported that Tβ4 upregulates antioxidative enzymes to protect human corneal epithelial cells against oxidative stress.[62]

We identified that under oxidative conditions Tβ4 significantly reduced ROS production and was associated with restoration of antioxidative enzymes and mitochondrial membrane potential (Fig. 3A).[63] Increased ROS further led to the loss of mitochondrial membrane potential ($\Delta\Psi m$), energy depletion, and, subsequently, increase in the Bax/Bcl$_2$ ratio favoring apoptosis.[64,65] Our observation further indicated that the loss of $\Delta\Psi m$ was prevented by pretreatment of Tβ4, providing a new role of Tβ4 in cardiac protection (Fig. 3B, last panel).

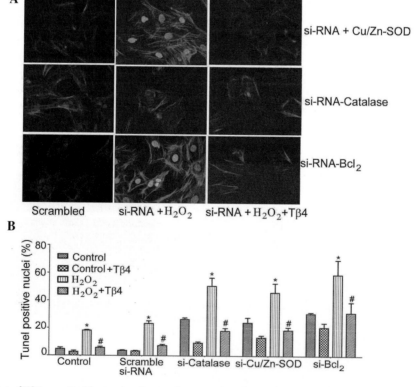

Figure 4. Effect of Tβ4 on antioxidant and antiapoptotic genes using siRNA of catalase, Cu/Zn-SOD, and Bcl$_2$ genes in cardiac fibroblasts. (A) Representative fluorescent microscopy images showing the effect of Tβ4 treatment in the presence and absence of H$_2$O$_2$-induced oxidative stress on cardiac fibroblasts transfected with siRNAs of Cu/Zn-SOD, catalase, and Bcl$_2$ versus scrambled siRNA, respectively. (B) Bar graph shows the percent TUNEL-positive nuclei under similar experimental conditions. Data represent the means ± SE of at least three separate experiments. A total of 65–82 nuclei were counted for each observation; *denotes $P <$ 0.05 compared to controls, while # denotes $P <$ 0.05, compared to the H$_2$O$_2$-treated group. Data in this figure are from Ref. 63.

Moreover, we noticed that Tβ4 reduced intracellular ROS by upregulating selected antioxidant genes, Cu/Zn-SOD, and catalase.[63] The mechanism of Cu/Zn-SOD and catalase upregulation by Tβ4 is currently unknown, but a transcription factor-mediation activity has been postulated.[66] Tβ4 has been reported to translocate into the nucleus possibly through its cluster of positively charged amino acid residues (KSKLKK), but the exact function is still obscure.[66] Oxidative stress is known to trigger apoptotic cell death;[67] our data further revealed that Tβ4 upregulated Bcl_2 (an antiapoptotic protein) and reduced excessive Bax expression and caspase-3 activation, thereby preventing apoptosis of cardiac fibroblasts from oxidative stress.[63] Most importantly, we found that Tβ4 reduced profibrotic gene expression in H_2O_2-treated cardiac fibroblasts,[63] profibrotic genes such as CTGF, collagen I, and collagen compared with H_2O_2-treated cells.[63] These findings corroborate our *in vivo* data.[36]

To further ascertain whether the effect of Tβ4 is mediated via antioxidant and antiapoptotic genes, we confirmed that Tβ4 failed to alleviate oxidative stress and prevent apoptosis when these molecules were knocked down in cells.[63] A TUNEL assay showed increased positive nuclei in H_2O_2-treated cells and in cells in which expression of catalase, Cu/Zn-SOD, and Bcl_2 were knocked down. Representative fluorescence microscopy images of TUNEL-positive nuclei (FITC-positive) are shown in Figure 4A. H_2O_2 treatment under similar experimental conditions showed an additive effect. Pretreatment with Tβ4 in the H_2O_2-treated group resulted in significant reduction in the TUNEL-positive nuclei (Fig. 4B). These results indicate that Tβ4 selectively targets antioxidant and antiapoptotic genes to provide cardiac protection under oxidative stress.

Conclusion

Our published studies have shown that treatment with Tβ4 significantly improves cardiac function in the MI model by suppressing NF-κB and thereby attenuating cardiac fibrosis, which offers a potential use of Tβ4 for therapeutic intervention. *In vitro* analyses further demonstrate that Tβ4 can reduce ROS accumulation by modulating the expression of anti-apoptotic and antioxidative genes and thereby preventing the loss of $\Delta\Psi m$ in oxidative stress. But while Tβ4 is cardio protective, it is not yet clear how Tβ4 exerts its beneficial effects. Although Tβ4 is internalized by cells, a specific cell surface receptor for Tβ4 has yet to be identified. Our results not only offer more mechanistic explanation about the protective role of Tβ4, they also call for further investigation of this peptide against oxidative damage in diverse pathologies where ROS plays a critical role. This makes Tβ4 a potential therapeutic target that may have wider applications once it is translated from the bench to bedside.

Conflicts of interest

The authors declare no conflicts of interest.

References

1. Bubb, M.R. 2003. Thymosin beta 4 interactions. *VitamHorm* **66:** 297–316.
2. Malinda, K.M., A.L. Goldstein & H.K. Kleinman. 1997. Thymosin beta 4 stimulates directional migration of human umbilical vein endothelial cells. *Faseb J.* **11:** 474–481.
3. Goldstein, A.L., E. Hannappel & H.K. Kleinman. 2005. Thymosin beta4: actin-sequestering protein moonlights to repair injured tissues. *Trends Mol. Med.* **11:** 421–429.
4. Hinkel, R. *et al.* 2010. Thymosin beta4: a key factor for protective effects of eEPCs in acute and chronic ischemia. *Ann. N. Y. Acad. Sci.* **1194:** 105–111.
5. Philp, D. *et al.* 2003. The actin binding site on thymosin beta4 promotes angiogenesis. *Faseb. J.* **17:** 2103–2105.
6. Low, T.L., S.K. Hu & A.L. Goldstein. 1981. Complete amino acid sequence of bovine thymosin beta 4: a thymic hormone that induces terminal deoxynucleotidyltransferase activity in thymocyte populations. *Proc. Natl. Acad. Sci. U. S. A.* **78:** 1162–1166.
7. Dedova, I.V. *et al.* 2006. Thymosin beta4 induces a conformational change in actin monomers. *Biophys. J.* **90:** 985–992.
8. Au, J.K., E.M. De La Cruz & D. Safer. 2007. Contributions from all over: widely distributed residues in thymosin beta-4 affect the kinetics and stability of actin binding. *Ann. N. Y. Acad. Sci.* **1112:** 38–44.
9. Goldstein, A.L. *et al.* 2012. Thymosin beta4: a multifunctional regenerative peptide. Basic properties and clinical applications. *Expert Opin. Biol. Ther.* **12:** 37–51.
10. Ehrlich, H.P. & S.W. Hazard, 3rd. 2010. Thymosin beta4 enhances repair by organizing connective tissue and preventing the appearance of myofibroblasts. *Ann. N. Y. Acad. Sci.* **1194:** 118–124.
11. Sosne, G. *et al.* 2006. Thymosin [beta]4 inhibits benzalkonium chloride-mediated apoptosis in corneal and conjunctival epithelial cells *in vitro*. *Exp. Eye Res.* **83:** 502–507.
12. Sosne, G. *et al.* 2010. Thymosin beta4 and corneal wound healing: visions of the future. *Ann. N. Y. Acad. Sci.* **1194:** 190–198.
13. Sosne, G., P. Qiu & M. Kurpakus-Wheater. 2007. Thymosin beta 4: a novel corneal wound healing and anti-inflammatory agent. *Clin. Ophthalmol.* **1:** 201–207.

14. Bock-Marquette, I. *et al.* 2004. Thymosin beta4 activates integrin-linked kinase and promotes cardiac cell migration, survival and cardiac repair. *Nature* **432:** 466–472.

15. Smart, N. *et al.* 2007. Thymosin beta4 induces adult epicardial progenitor mobilization and neovascularization. *Nature* **445:** 177–182.

16. Smart, N. *et al.* 2010. Thymosin beta4 facilitates epicardial neovascularization of the injured adult heart. *Ann. N. Y. Acad. Sci.* **1194:** 97–104.

17. Dube, K.N. *et al.* 2012. Thymosin beta4 protein therapy for cardiac repair. *Curr. Pharm. Des.* **18:** 799–806.

18. Kupatt, C., I. Bock-Marquette & P. Boekstegers. 2008. Embryonic endothelial progenitor cell-mediated cardioprotection requires Thymosin beta4. *Trends Cardiovasc. Med.* **18:** 205–210.

19. Qiu, F.Y. *et al.* 2009. Thymosin beta4 induces endothelial progenitor cell migration via PI3K/Akt/eNOS signal transduction pathway. *J. Cardiovasc. Pharmacol.* **53:** 209–214.

20. Bock-Marquette, I. *et al.* 2009. Thymosin beta4 mediated PKC activation is essential to initiate the embryonic coronary developmental program and epicardial progenitor cell activation in adult mice *in vivo*. *J. Mol. Cell Cardiol.* **46:** 728–738.

21. Swynghedauw, B. 1999. Molecular mechanisms of myocardial remodeling. *Physiol. Rev.* **79:** 215–262.

22. Gonzalez, A. *et al.* 2011. New targets to treat the structural remodeling of the myocardium. *J. Am. Coll. Cardiol.* **58:** 1833–1843.

23. Srivastava, D. *et al.* 2007. Thymosin beta4 is cardioprotective after myocardial infarction. *Ann. N. Y. Acad. Sci.* **1112:** 161–170.

24. Shrivastava, S. *et al.* 2010. Thymosin beta4 and cardiac repair. *Ann. N. Y. Acad. Sci.* **1194:** 87–96.

25. Crockford, D. 2007. Development of thymosin beta4 for treatment of patients with ischemic heart disease. *Ann. N. Y. Acad. Sci.* **1112:** 385–395.

26. Yesilay, A.B. *et al.* 2011. Thymosin beta4 levels after successful primary percutaneous coronary intervention for acute myocardial infarction. *Turk KardiyolojiDernegiarsivi: Turk KardiyolojiDernegininyayinorganidir.* **39:** 654–660.

27. Lu, H. *et al.* 2006. Integrin-linked kinase expression is elevated in human cardiac hypertrophy and induces hypertrophy in transgenic mice. *Circulation* **114:** 2271–2279.

28. White, D.E. *et al.* 2006. Targeted ablation of ILK from the murine heart results in dilated cardiomyopathy and spontaneous heart failure. *Genes Dev.* **20:** 2355–2360.

29. Hannigan, G.E., J.G. Coles & S. Dedhar. 2007. Integrin-linked kinase at the heart of cardiac contractility, repair, and disease. *Circ. Res.* **100:** 1408–1414.

30. Knoll, R. *et al.* 2007. Laminin-alpha4 and integrin-linked kinase mutations cause human cardiomyopathy via simultaneous defects in cardiomyocytes and endothelial cells. *Circulation* **116:** 515–525.

31. Ding, L. *et al.* 2009. Increased expression of integrin-linked kinase attenuates left ventricular remodeling and improves cardiac function after myocardial infarction. *Circulation* **120:** 764–773.

32. Gu, R. *et al.* 2012. Increased expression of integrin-linked kinase improves cardiac function and decreases mortality in dilated cardiomyopathy model of rats. *PLoS One* **7:** e31279.

33. Legate, K.R., S.A. Wickstrom & R. Fassler. 2009. Genetic and cell biological analysis of integrin outside-in signaling. *Genes. Dev.* **23:** 397–418.

34. Fukuda, T. *et al.* 2003. PINCH-1 is an obligate partner of integrin-linked kinase (ILK) functioning in cell shape modulation, motility, and survival. *J. Biol. Chem.* **278:** 51324–51333.

35. Fukuda, T. *et al.* 2003. CH-ILKBP regulates cell survival by facilitating the membrane translocation of protein kinase B/Akt. *J. Cell Biol.* **160:** 1001–1008.

36. Sopko, N. *et al.* 2011. Significance of thymosin beta4 and implication of PINCH-1-ILK-alpha-Parvin (PIP) complex in human dilated cardiomyopathy. *PLoS One* **6:** e20184.

37. Hinkel, R. *et al.* 2008. Thymosin beta4 is an essential paracrine factor of embryonic endothelial progenitor cell-mediated cardioprotection. *Circulation* **117:** 2232–2240.

38. Zhou, B. *et al.* 2012. Thymosin beta 4 treatment after myocardial infarction does not reprogram epicardial cells into cardiomyocytes. *J. Mol. Cell Cardiol.* **52:** 43–47.

39. Martinon, F., A. Mayor & J. Tschopp. 2009. The inflammasomes: guardians of the body. *Annu. Rev. Immunol.* **27:** 229–265.

40. Banerjee, I. *et al.* 2012. Thymosin beta 4 is dispensable for murine cardiac development and function. *Circ. Res.* **110:** 456–464.

41. Yellon, D.M. & D.J. Hausenloy. 2007. Myocardial reperfusion injury. *N. Engl. J. Med.* **357:** 1121–1135.

42. Wrigley, B.J., G.Y. Lip & E. Shantsila. 2011. The role of monocytes and inflammation in the pathophysiology of heart failure. *Eur. J. Heart Fail.* **13:** 1161–1171.

43. Badamchian, M. *et al.* 2003. Thymosinbeta(4) reduces lethality and down-regulates inflammatory mediators in endotoxin-induced septic shock. *Int. Immunopharmacol.* **3:** 1225–1233.

44. Cavasin, M.A. 2006. Therapeutic potential of thymosin-beta4 and its derivative N-acetyl-seryl-aspartyl-lysyl-proline (Ac-SDKP) in cardiac healing after infarction. *Am. J. Cardiovasc. Drugs* **6:** 305–311.

45. Girardi, M. *et al.* 2003. Anti-inflammatory effects in the skin of thymosin-beta4 splice-variants. *Immunology* **109:** 1–7.

46. Sosne, G. *et al.* 2002. Thymosin beta 4 promotes corneal wound healing and decreases inflammation *in vivo* following alkali injury. *Exp. Eye Res.* **74:** 293–299.

47. Sosne, G. *et al.* 2005. Thymosin-beta4 modulates corneal matrix metalloproteinase levels and polymorphonuclear cell infiltration after alkali injury. *Invest. Ophthalmol. Vis. Sci.* **46:** 2388–2395.

48. Gutierrez, S.H., M.R. Kuri & E.R. del Castillo. 2008. Cardiac role of the transcription factor NF-kappaB. *Cardiovasc. Hematol. Disord. Drug Targets* **8:** 153–160.

49. Gordon, J.W., J.A. Shaw & L.A. Kirshenbaum. 2011. Multiple facets of NF-kappaB in the heart: to be or not to NF-kappaB. *Circ. Res.* **108:** 1122–1132.

50. Sosne, G. *et al.* 2007. Thymosin beta 4 suppression of corneal NFkappaB: a potential anti-inflammatory pathway. *Exp. Eye Res.* **84:** 663–669.

51. Qiu, P. *et al.* 2011. Thymosin beta4 inhibits TNF-alpha-induced NF-kappaB activation, IL-8 expression, and the sensitizing effects by its partners PINCH-1 and ILK. *Faseb J.* **25:** 1815–1826.

52. Creemers, E.E. & Y.M. Pinto. 2011. Molecular mechanisms that control interstitial fibrosis in the pressure-overloaded heart. *Cardiovasc. Res.* **89:** 265–272.

53. Krenning, G., E.M. Zeisberg & R. Kalluri. 2010. The origin of fibroblasts and mechanism of cardiac fibrosis. *J. Cell. Physiol.* **225:** 631–637.

54. Espira, L. & M.P. Czubryt. 2009. Emerging concepts in cardiac matrix biology. *Can. J. Physiol. Pharmacol.* **87:** 996–1008.

55. Yang, F. *et al.* 2004. Ac-SDKP reverses inflammation and fibrosis in rats with heart failure after myocardial infarction. *Hypertension* **43:** 229–236.

56. Cavasin, M.A. *et al.* 2007. Decreased endogenous levels of Ac-SDKP promote organ fibrosis. *Hypertension* **50:** 130–136.

57. Tsutsui, H., S. Kinugawa & S. Matsushima. 2011. Oxidative stress and heart failure. *Am J. Physiol. Heart Circ. Physiol.* **301:** H2181–H2190.

58. Afanas'ev, I. 2011. ROS and RNS signaling in heart disorders: could antioxidant treatment be successful? *Oxid. Med. Cell. Longev.* **2011:** 293769.

59. Giordano, F.J. 2005. Oxygen, oxidative stress, hypoxia, and heart failure. *J. Clin. Invest.* **115:** 500–508.

60. Ide, T. *et al.* 1999. Mitochondrial electron transport complex I is a potential source of oxygen free radicals in the failing myocardium. *Circ. Res.* **85:** 357–363.

61. Murphy, M.P. 2009. How mitochondria produce reactive oxygen species. *Biochem. J.* **417:** 1–13.

62. Ho, J.H. *et al.* 2008. Thymosin beta-4 upregulates anti-oxidative enzymes and protects human cornea epithelial cells against oxidative damage. *Br. J. Ophthalmol.* **92:** 992–997.

63. Kumar, S. & S. Gupta. 2011. Thymosin beta 4 prevents oxidative stress by targeting antioxidant and anti-apoptotic genes in cardiac fibroblasts. *PLoS One* **6:** e26912.

64. Weiss, J.N. *et al.* 2003. Role of the mitochondrial permeability transition in myocardial disease. *Circ. Res.* **93:** 292–301.

65. Dai, D.F. *et al.* 2012. Mitochondrial proteome remodelling in pressure overload-induced heart failure: the role of mitochondrial oxidative stress. *Cardiovasc. Res.* **93:** 79–88.

66. Huff, T. *et al.* 2004. Nuclear localisation of the G-actin sequestering peptide thymosin beta4. *J. Cell Sci.* **117:** 5333–5341.

67. Hori, M. & K. Nishida. 2009. Oxidative stress and left ventricular remodelling after myocardial infarction. *Cardiovasc. Res.* **81:** 457–464.

Ann. N.Y. Acad. Sci. ISSN 0077-8923

ANNALS OF THE NEW YORK ACADEMY OF SCIENCES

Issue: *Thymosins in Health and Disease*

Myocardial regeneration: expanding the repertoire of thymosin β4 in the ischemic heart

Nicola Smart,[1] Sveva Bollini,[1] Karina N. Dubé,[1] Joaquim M. Vieira,[1] Bin Zhou,[2,3] Johannes Riegler,[4,5] Anthony N. Price,[6] Mark F. Lythgoe,[4] Sean Davidson,[7] Derek Yellon,[7] William T. Pu,[2,3] and Paul R. Riley[1]

[1]Department of Physiology, Anatomy, and Genetics, University of Oxford, Oxford, United Kingdom. [2]Harvard Stem Cell Institute and Department of Cardiology, Children's Hospital Boston, Boston, Massachusetts. [3]Department of Genetics, Harvard Medical School, Boston, Massachusetts. [4]Center for Advanced Biomedical Imaging (CABI), Department of Medicine and Institute of Child Health, University College London (UCL), London, United Kingdom. [5]Centre for Mathematics and Physics in the Life Sciences and Experimental Biology (CoMPLEX), UCL, London, United Kingdom. [6]MRC Clinical Sciences Center, Faculty of Medicine, Imperial College London, London, United Kingdom. [7]The Hatter Cardiovascular Institute, University College London, London, United Kingdom

Address for correspondence: Paul Riley, Department of Physiology, Anatomy, and Genetics, Sherrington Building, University of Oxford, South Parks Road, Oxford, UK, OX1 3PT. paul.riley@dpag.ox.ac.uk

Efficient cardiac regeneration postinfarction (MI) requires the replacement of lost cardiomyocytes, formation of new coronary vessels and appropriate modulation of the inflammatory response. However, insight into how to stimulate repair of the human heart is currently limited. Using the embryonic paradigm of regeneration, we demonstrated that the actin-binding peptide thymosin β4 (Tβ4), required for epicardium-derived coronary vasculogenesis, can recapitulate its embryonic role and activate quiescent adult epicardial cells (EPDCs). Once stimulated, EPDCs facilitate neovascularization of the ischemic adult heart and, moreover, contribute bona fide cardiomyocytes. EPDC-derived cardiomyocytes structurally and functionally integrate with resident muscle to regenerate functional myocardium, limiting pathological remodeling, and effecting an improvement in cardiac function. Alongside pro-survival and anti-inflammatory properties, these regenerative roles, via EPDCs, markedly expand the range of therapeutic benefits of Tβ4 to sustain and repair the myocardium after ischemic damage.

Keywords: thymosin β4; EPDCs; epicardium; *Wt1*; myocardial regeneration; *de novo* cardiomyocytes

The quest for effective cardiac regenerative strategies

Cardiovascular diseases (CVD) are the leading cause of morbidity and mortality worldwide, with myocardial infarction (MI) being the most common cause of cardiac injury. The consequence of losing over a billion functional cardiomyocytes in mammals is replacement by a noncontractile scar, pathological remodeling, and progression to heart failure (HF).[1] Since the mammalian heart is unable to adequately regenerate beyond early postnatal stages,[2] transplantation is the only possible cure, although confounded by host immune rejection and a limited supply of donor hearts. Conventional palliative medication seeks to preserve the already compromised cardiac function, rather than regenerating lost myocardium. Consequently, there has been an intensive effort to develop stem cell–based strategies for cardiac repair to both regenerate heart muscle and promote coronary vasculogenesis after MI, culminating in a number of clinical trials. The majority of such trials have relied on the use of autologous cell types, including bone marrow stem cells; however, there is no consensus that these cells can differentiate to contribute new muscle efficiently. Perhaps for this reason, the clinical improvement has been generally disappointing.[3] Hence, there continues to be an urgent need to identify the most promising cardiovascular stem cells; either for transplantation and engraftment or activation of a resident population for induction toward a cardiomyocyte fate (reviewed in Ref. 4).

doi: 10.1111/j.1749-6632.2012.06708.x

Redeploying embryonic mechanisms in the adult heart for repair

A contemporary paradigm in regenerative medicine is that tissue repair in the adult is frequently underpinned by a reactivation of the embryonic program that created the tissue in the first instance. As such, there is much to gain from understanding the embryonic mechanisms of vasculogenesis and cardiogenesis. The success of this approach lies in the identification of a tractable progenitor cell population and the development of appropriate strategies for their redeployment in the adult, based on defined embryonic roles. Due to their fundamental role in heart development, the epicardium-derived cells (EPDCs) have emerged as a population that fulfill this remit and have come under intense scrutiny as a new source for myocardial regeneration. The epicardium gives rise to EPDCs, multipotent cardiac progenitors that were proposed to be true cardiac stem cells,[5] due to their potential to differentiate into endothelial cells (ECs)[6], coronary vascular smooth muscle cells (VSMCs)[7] cardiac fibroblasts,[8] and cardiomyocytes.[5] In addition to direct contribution of cardiac cells, the epicardium plays a critical role via paracrine secretion of key signaling factors in myocardial compaction, Purkinje fiber development and inhibition of endocardial epithelial–mesenchymal transition (EMT).[9]

Thymosin β4 reactivates the quiescent adult epicardium

A prerequisite for using EPDCs for repair in the adult heart is the identification of factors to reactivate this ordinarily dormant reservoir of cells in order to exploit their restorative power. Valuable insight into the epicardial response to injury may be derived from studies in zebrafish. Following resection of the adult fish heart, the epicardium exhibits a rapid and robust response to injury, which includes the reexpression of embryonic epicardial markers, *Tbx18* and *Raldh2*, and proliferation of EPDCs within one to two days of resection.[10] Lacking the intrinsic regenerative capacity of fish, the mammalian epicardium requires a boost to achieve even a modest improvement in myocardial regeneration. Reactivation of the mammalian adult epicardium was first revealed using the actin-monomer—binding protein thymosin β4 (Tβ4).[11] Tβ4-treated, infarcted hearts re-

vealed dramatic EPDC proliferation and large numbers of ECs and VSMCs in the expanded subepicardial space, which assembled to form a capillary network. Myocardial injury is itself sufficient to promote epicardial activation and neovascularisation, and Tβ4 appears to act synergistically to augment the extent of repair.[12] Compared with a modest degree of VSMC recruitment in vehicle-treated hearts, extensive VSMC migration and differentiation and a significant increase in the number of perfused, functional VSMC-lined arterioles was observed following Tβ4 treatment. Arteriogenesis may therefore explain the beneficial effects of Tβ4 treatment post-MI, beyond the relatively unstable and grossly inadequate endogenous capillary response.

Thymosin β4 and EPDCs: a joint force for myocardial regeneration

Just as the myocardium and coronary vasculature develop simultaneously in a coordinated manner in the embryo, it is highly desirable, therapeutically, to reinstate all damaged components concurrently within the diseased adult heart. Indeed, attempts to replenish muscle in an ischemic environment are futile. The epicardium, by contributing to the myocardium and coronary vasculature during development, offers a unique target for coordinately stimulating myocardial and coronary vascular repair. Moreover, epicardial thickening in response to MI has recently been shown to act as a source of trophic paracrine factors that condition the underlying myocardium for repair in an analogous manner to nurturing myocardial growth during development.[13]

From the demonstration of a significant epicardial contribution to the cardiomyocyte lineage, a basis emerged for translating myocardial potential into the adult heart.[14] However, epicardial lineage tracing is confounded in the adult since expression of most known epicardial genes is restricted to embryonic stages and silenced in adulthood. From a peak of expression at E11.5, coincident with epicardial formation and EPDC contribution, *Wt1*, alongside other epicardial genes, is downregulated by early postnatal stages to leave the adult epicardium effectively dormant (Fig. 1A). We exploited the ability of Tβ4 to reactivate the adult epicardium in order to restore *Wt1* expression (Fig. 1B). "Priming" of adult Wt1[GFPCre/+] and Wt1[CreERT2]; R26R[EYFP/+] knockin mice with Tβ4 in this manner enabled both constitutive (GFP+) and pulse- (YFP+) labeling

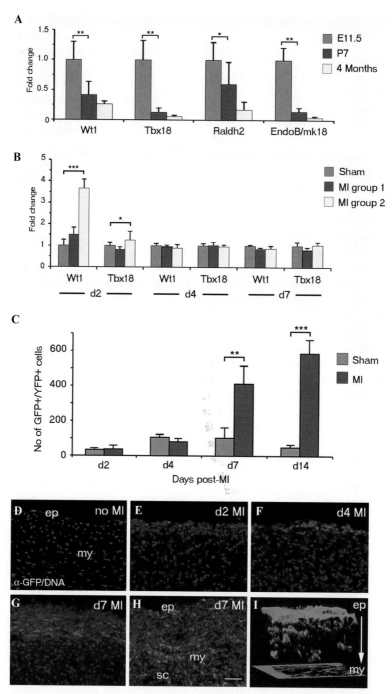

Figure 1. Myocardial injury and Tβ4 reactivate embryonic epicardial gene expression in the adult heart. Real-time qRT-PCR revealed a significant downregulation ($^*P \leq 0.05$; $^{**}P \leq 0.01$) of the epicardial genes *Wt1*, *Tbx18*, *Raldh2* and *EndoB* in intact hearts from embryonic day E11.5 to postnatal day 7 (P7) and four months of age (A). Tβ4 priming resulted in a precocious upregulation in *Wt1* ($^{***}P \leq 0.001$) and *Tbx18* ($^*P \leq 0.05$) as early as day 2 post-MI, the extent of which correlated with severity of injury. MI group 1 = mild injury; MI group 2 = severe injury (B). An increase in the number of EPDCs was observed following injury from days 2, 4, 7 ($^{**}P \leq 0.01$) to day 14 ($^{***}P \leq 0.00$; C–G), with evidence of migration into the underlying myocardium (H–I). Multiphoton-acquired 3D reconstructions revealed the extent of EPDC migration from the epicardial surface to the underlying myocardium in response to injury and Tβ4 (I). Scale bar in H represents 50 μm (applies to D–H). ep, epicardium; my, myocardium; sc, scar region. All statistics: Student's *t*-test. Data are from Ref. 15.

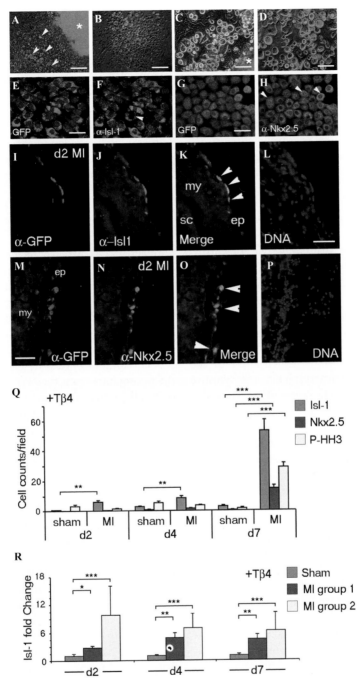

Figure 2. Tβ4-primed EPDCs give rise to cardiac progenitors. Primed GFP⁺ EPDCs (white arrowheads) migrated from epicardial explants (A–D). GFP⁺ EPDCs costrained for Isl-1 (E and F); white arrowhead indicates an Isl-1⁻ EPDC surrounded by an Isl-1⁺ cluster) or Nkx2.5 (G and H; white arrowheads indicate Nkx2.5⁺ EPDCs) providing evidence of cardiac progenitor formation. *In vivo*, YFP⁺ EPDCs coexpressing Isl-1 (I–L) or Nkx2.5 (M–P) resided in the epicardium, indicative of an early progenitor response two days post-MI. Significant increases in Isl-1⁺/YFP⁺ cells and Nkx2.5⁺/YFP⁺ cells were observed within seven days post-MI, alongside increased phospho-histone H3⁺ (P-HH3⁺) proliferating YFP⁺ progenitors at day 7 compared to sham-operated controls (Q). Real-time qRT-PCR revealed a significant increase in Isl-1 expression in primed hearts following MI correlating with the extent of injury (R; * $P \le 0.05$; ** $P \le 0.01$; *** $P \le 0.001$; MI group 1 = mild injury; group 2 = severe injury). All *P* values calculated by the paired ANOVA (Q) and Student's *t*-test (R). Scale bars in A, 100 μm; B, 500 μm; C–E, G, L, M, 50 μm. ep, epicardium; my, myocardium; sc, scar region. Data are from Ref. 15.

of EPDCs, respectively, to facilitate the tracking of epicardium-derived precursors through to mature cardiomyocytes.[15] Proliferative YFP$^+$ cells were shown to expand in number over time (d2–d14) within the epicardium, extending into the myocardium, toward the region of injury (Fig. 1C–H). Two-photon molecular excitation laser scanning microscopy revealed live YFP$^+$ cells in the epicardium and subepicardial region at day 7 post-MI, and a large number of cells that had migrated into the underlying myocardiumto a depth of ≤ 100 μm (Fig 1I).

In epicardial explants[16] from Tβ4-primed Wt1$^{GFPCre/+}$ adult hearts, a proportion of outgrowing GFP$^+$ cells were positive for Isl-1 and Nkx2.5, indicative of cardiomyocyte precursors (Fig. 2A–H).[17] After 14 days of culture, cells adopted a more differentiated cardiac muscle phenotype, with expression of sarcomeric structural components, cardiac troponin T (cTnT), sarcomeric α-actinin (SαA), and cardiac myosin–binding protein C (cMyBPC).

In an *in vivo* model of MI (ligation of the left anterior coronary artery), YFP$^+$ cells positive for Isl-1 or Nkx2.5 cells were found to reside in the epicardium and adjacent subepicardial region of Tβ4 -treated Wt1^{CreERT2}; R26R$^{EYFP/+}$ hearts within two days post-MI (Fig. 2I–Q). This was accompanied by elevated *Isl-1* expression from two days post-MI onward (Fig. 2R). By day 14 post-MI, larger YFP$^+$ cells, coexpressing sarcomeric markers and morphologically resembling mature cardiomyocytes, were detected within the left ventricular wall, specifically in the border zone and peri-infarct region (Fig. 3 A–H). The total mean percentage of YFP$^+$ EPDCs that became cardiomyocytes was 0.59 ± 0.18% (serial sections through $n = 7$ hearts $\pm$ SEM). Approximately 82% of YFP$^+$ cardiomyoctes were located proximal to the site of injury, residing either in the border zone or within the immediate surrounding healthy myocardium. Consistent with tracking-labeled cardiomyocytes from EPDCs, BrdU injections of Wt1-ERT2YFP mice revealed proliferating BrdU$^+$ EPDCs at day 4 traced to BrdU$^+$/YFP$^+$ cardiomyocytes at day 14 (data not shown).[15] Importantly, the *de novo* cardiomyocytes were appropriately integrated with resident myocardium and with each other, via both adherens (Fig. 3D–E) and gap junction formation (Fig. 3G). Moreover, functional integration of YFP$^+$ cardiomy

ocytes with existing myocardium was demonstrated upon measuring $[Ca^{2+}]_i$ transients with neighboring YFP$^-$ cardiomyocytes. We recorded evoked cellular calcium transients $[Ca^{2+}]_i$ between YFP$^+$ and YFP$^-$ cells in Langendorff-perfused, Tβ4-primed, Wt1-ERT2YFP hearts loaded with the fluorescent indicator rhod-2/AM (Fig. 3I–N). We observed small, immature YFP$^+$ rod-shaped cells within the border zone that failed to elicit $[Ca^{2+}]_i$ transients, as well as mature YFP$^+$ cardiomyocytes in which calcium transients were synchronous with kinetics indistinguishable from those of neighboring YFP$^-$ cardiomyocytes (Fig. 3O). Minor differences in resolution between resident YFP$^-$ and the *de novo* YFP$^+$ cardiomyocyte transients (compare Fig. 3M with Fig. 3N) likely reflect the newly acquired function of the YFP$^+$ population. These results strongly suggest that Tβ4-primed EPDC-derivatives couple with surviving cardiomyocytes following injury to form a functional syncytium with resident myocardium.

Excluding the potential for inadvertent tracing of resident cardiomyocytes due to ectopic activation of the fluorophore from the *Wt1* knock-in alleles required the transplantation of FACS-isolated donor GFP$^+$ cells into nontransgenic host hearts (data not shown).[15] Extensive analysis of the prospective donor cells, isolated at day 4, characterized them as EPDCs and ruled out any coisolation of nonepicardial cells that may have upregulated *Wt1*. Immunostaining confirmed restricted localization of Wt1$^+$ cells to the epicardium and expanded subepicardial region. GFP$^+$ cells were excluded from the entire myocardium and no cardiomyocyte marker expression was detectable from the FACS-isolated donor cells at day 4. These data collectively confirmed that donor Wt1$^+$ cells were exclusively epicardial and were not myocytes that had simply upregulated GFP via the targeted *Wt1* allele, consistent with the fact that GFP$^+$ or YFP$^+$ cardiomyocytes were never observed at day 4 post-MI. FACS-isolated donor cells from Tβ4/injury-primed male or female Wt1-GFP animals (between 3×10^4 and 6×10^4 cells/host) were injected into the subepicardial space of Tβ4 primed nontransgenic gender-mismatched hosts, following LAD ligation. At 24 h following transplantation, small round GFP$^+$ cells were restricted to the site of injection within the host heart. By day 7, GFP$^+$ cells residing within the subepicardium were Nkx2.5$^+$, indicative of myocardial progenitor

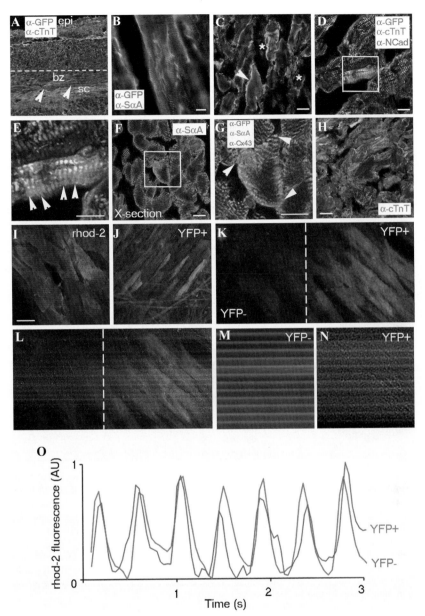

Figure 3. Activated adult Wt1[+] EPDCs differentiate into structurally coupled, functional cardiomyocytes. YFP[+] cells (detected by α-GFP), coexpressing cTnT, residing within the border zone and peri-infarct region (A; white arrowheads; dashed white line demarcates extent of peri-infarct region). The YFP[+] cells had evident sarcomeric structure, which were SαA[+] (B, C, F, G) or cTNT[+] (D, E, H), with sarcomeric banding (white arrowheads) and evidence of structural coupling to resident YFP[−] cardiomyocytes through N-Cad[+] adherens junction (D and E) and Cx43[+] gap junction formation (F and G). Multiphoton imaging on Langendorff-perfused Tβ4-primed Wt1-ERT2YFP hearts loaded with rhod-2/AM at day 14 post-MI. Rhod-2 loading (I) of YFP[+] cardiomyocytes within the peri-infarct region (J) was compared against distal YFP[−] cardiomyocytes (K). EYFP and rhod-2/AM excited simultaneously at 990 nm allowed measurement of calcium transients ($[Ca^{2+}]_i$ line traces) across clustered YFP[−] and YFP[+] cardiomyocytes (l) as evidence of functional coupling both between YFP[−] with YFP[+] and YFP[+] with YFP[+] cardiomyocytes. Distal YFP[−] spontaneous calcium transients were visualized without electrical pacing (M) and compared against YFP[+] transients (N). YFP[−] and YFP[+] calcium transients displayed comparable amplitude and synchrony. Scale bars represent 150 μm (A); 20 μm (B–H); bz, border zone; sc, scar region. Data are from Ref. 15.

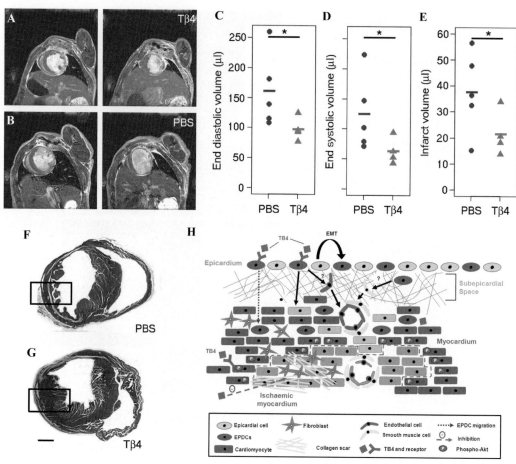

Figure 4. EPDC-derived cardiomyocytes contribute toward improved cardiac function and myocardial regeneration post-MI. MRI analyses with representative short axis images (mid-ventricular and apical slices) of infarcted hearts 28 days post-MI with late gadolinium enhancement (scar tissue bright) following treatment with either Tβ4 (A) or PBS (B). Scatter plots for end diastolic volume (EDV; C), end systolic volume (ESV; D), and infarct volume (E) indicate improved cardiac function following Tβ4 treatment. Horizontal bars indicate the means for each group. Trichrome-stained transverse sections from PBS− (F) and Tβ4− (G)-treated hearts 28 days post-MI revealed reduced scarring and fibrosis with Tβ4 treatment (blue represents collagen deposition). Note the increased proportion of healthy myocardium highlighted by the black box in (G) relative to the comparative region in (F). * $P \leq 0.05$; all statistics by one-way repeated measures ANOVA. Scale bar in G represents 1 mm (applies to F, G). Myocardial replenishment constitutes one of the beneficial effects of Tβ4 in the ischemic heart, alongside promotion of cell survival and neovascularization and modulation of the inflammatory response (H). Data are from Ref. 15.

commitment, while other GFP+ cells in the underlying myocardium, although morphologically immature, were found to coexpress cTnT. More definitive donor GFP+ cardiomyocytes with myofibrillar structure coexpressing cTnT were observed within the host myocardium by day 14. Collectively, the detection of these staged donor derivatives suggested progressive differentiation toward a mature cardiomyocyte fate. Fluorescence *in situ* hybridization of GFP+ cardiomyocytes confirmed a single XX or XY karyotype in female or male-derived donor

cells, respectively, implying transdifferentiation of EPDCs in the absence of cell fusion.[15]

MRI analyses over a 28-day time course after MI revealed that Tβ4 treatment resulted in significant improvement in ejection fraction and end diastolic/systolic volumes over time (Fig. 4A–E). Improvements in LV mass, relative to scar volume (Fig. 4F–G), suggested a degree of muscle regeneration attributable, in part, to the primed EPDC contribution of functional cardiomyocytes.

Therapeutic application of Tβ4 for the ischemic heart

Despite the beneficial outcome, a limitation of this approach is the requirement to pretreat with Tβ4 prior to injury which, from a translational perspective, would require the prior identification of patients considered at risk of MI. No cardiomyocyte replenishment from epicardial precursors was achieved without pretreatment [13,15] or when Tβ4 was administered only after MI.[18] Whether epicardial activation prior to injury is required to alter the subsequent fate of epicardial cells or whether it simply enhances the magnitude of epicardial reactivation sufficiently, over and above the activation that occurs by posttreatment with Tβ4 remains to be determined. In addition, the incidence of Tβ4-induced cardiomyocytes from activated EPDCs was very low and certainly insufficient to replenish losses in the region of 10^9 cardiomyocytes post-MI; however, the potential myocardial regeneration via the epicardium paves the way for the discovery of novel compounds that more efficiently promote regeneration without a requirement to pretreat. Replenishment of destroyed myocardium by activated EPDCs is a significant advance toward resident cell-based therapy for acute MI in human patients. Thus, in addition to its prosurvival effects,[19] Tβ4 can activate adult epicardium to not only induce neovascularization[12] but promote myocardial regeneration[15] to restore functional vasculature, maintain cardiomyocyte survival, and replace lost muscle in the injured heart.

Pleiotropic roles of Tβ4 in cardiac repair

The direct reactivation of adult EPDCs by Tβ4 to underpin the initiation and migration of adult epicardium-derived cardiovascular progenitors toward a cardiomyocyte fate constitutes one of several pleiotropic benefits of Tβ4 therapy for acute myocardial infarction (Fig. 4H). Previously, Tβ4 was shown to activate Akt signaling to enhance cardiomyocyte survival postischemic injury.[19] In addition, we and others have revealed that Tβ4 can induce the adult epicardium in the same setting to contribute coronary endothelial and smooth muscle cells and initiate vascular repair.[12,20]

Alongside neovascularization and myocardial regeneration, efficient therapies to combat cardiac dysfunction also require appropriate modulation of inflammatory responses to prevent maladaptive remodeling and fibrosis.[21] Following injury, local production of cytokines instructs recruitment of phagocytes for the removal of necrotic debris and macrophages to promote the formation of granulation tissue.[22] Uncontrolled inflammation, as manifested post-MI, inhibits wound healing and manipulation of the early response presents a therapeutic target to salvage viable myocardium and improve repair. Tβ4 is markedly upregulated around the sites of inflammation[23] and exhibits anti-inflammatory properties in many pathological settings.[24–28] Proinflammatory cytokines produced during MI, including IL-1β and TNF-α, converge to stimulate activation and nuclear transport of the transcription factor NF-κB. Chronic activation of NF-κB in the ischemic heart, exacerbates cardiac remodeling by imparting proinflammatory, profibrotic, and proapoptotic effects.[29] In models of MI and heart failure, Tβ4 was shown to suppress NF-κB activation and collagen expression— hallmarks of cardiac fibrosis.[30] While the mechanism of Tβ4-induced NF-κB suppression has not been defined, Tβ4 modulates the PINCH-ILK complex, the activation of which is known to sensitize NF-κB signaling after TNF-α stimulation in acute inflammatory conditions,[31] suggesting a possible involvement of this pathway. Further antifibrotic properties of Tβ4 may be mediated via the biological activity of its tetrapeptide cleavage product, AcSDKP, which inhibits collagen deposition by cardiac fibroblasts in the failing heart.[32]

Prospects for further extending the limits of the therapeutic utility of Tβ4 are emerging with the recent demonstration that Tβ4 enhances *in vivo* reprogramming of cardiac fibroblasts into induced cardiomyocytes (iCMs) upon viral transduction with the transcription factors Gata4, Mef2c, and Tbx5.[33] By activating EPDCs, thereby increasing the number and proliferation of cardiac fibroblasts, Tβ4 treatment enhanced the yield of iCMs upon transduction.

The multiple facets of myocardial repair addressed by delivery of a single peptide distinguish Tβ4 as an attractive therapeutic agent, particularly given its intrinsic capacity to generate potent anti-inflammatory and antifibrotic derivatives such as AcSDKP. Further contributions via the epicardium to regenerate lost myocardium and restore blood flow to the ischemic heart expand its therapeutic repertoire immeasurably. The multiple beneficial facets of Tβ4 are delineated with time post-MI. The prosurvival and anti-inflammatory activities of Tβ4 are early injury responses, to maintain the quotient of surviving myocardium, while neovascularization and *de novo* cardiogenesis are longer-term regenerative functions acting through the common target of adult EPDCs.

Acknowledgments

This work was funded by the British Heart Foundation.

Conflicts of interest

The authors declare no conflicts of interest.

References

1. Reinecke, H., E. Minami, W.Z. Zhu & M.A. Laflamme. 2008. Cardiogenic differentiation and transdifferentiation of progenitor cells. *Circ. Res.* **103:** 1058–1071.
2. Porrello, E.R., A.I. Mahmoud, E. Simpson, *et al.* 2011. Transient regenerative potential of the neonatal mouse heart. *Science* **331:** 1078–1080.
3. Bartunek, J., S. Dimmeler, H. Drexler, *et al.* 2006. The consensus of the task force of the European Society of Cardiology concerning the clinical investigation of the use of autologous adult stem cells for repair of the heart. *Eur. Heart J.* **27:** 1338–1340.
4. Laflamme, M.A. & C.E. Murry. 2011. Heart regeneration. *Nature* **473:** 326–335.
5. Wessels, A. & J.M. Perez-Pomares. 2004. The epicardium and epicardially derived cells (EPDCs) as cardiac stem cells. *Anat. Rec. A Discov. Mol. Cell Evol. Biol.* **276:** 43–57.
6. Perez-Pomares, J.M., R. Carmona, M. Gonzalez-Iriarte, *et al.* 2002. Origin of coronary endothelial cells from epicardial mesothelium in avian embryos. *Int. J. Dev. Biol.* **46:** 1005–1013.
7. Mikawa, T. & R.G. Gourdie. 1996. Pericardial mesoderm generates a population of coronary smooth muscle cells migrating into the heart along with ingrowth of the epicardial organ. *Developmental Biol.* **174:** 221–232.
8. Gittenberger-De Groot, A.C., M.P. Vrancken Peeters, M.M.T. Mentink, *et al.* 1998. Epicardium-derived cells contribute a novel population to the myocardial wall and the atrioventricular cushions. *Circulation Res.* **82:** 1043–1052.
9. Winter, E.M. & A.C. Gittenberger-De Groot. 2007. Epicardium-derived cells in cardiogenesis and cardiac regeneration. *Cell Mol. Life Sci.* **64:** 692–703.
10. Lepilina, A., A.N. Coon, K. Kikuchi, *et al.* 2006. A dynamic epicardial injury response supports progenitor cell activity during zebrafish heart regeneration. *Cell* **127:** 607–619.
11. Smart, N., C.A. Risebro, A.A.D. Melville, *et al.* 2007. Thymosin β4 induces adult epicardial progenitor mobilization and neovascularization. *Nature* **445:** 177–182.
12. Smart, N., C.A. Risebro, J.E. Clark, *et al.* 2010. Thymosin beta4 facilitates epicardial neovascularization of the injured adult heart. *Ann. N.Y. Acad. Sci.* **1194:** 97–104.
13. Zhou, B., L.B. Honor, H. He, *et al.* 2011. Adult mouse epicardium modulates myocardial injury by secreting paracrine factors. *J. Clin. Invest* **121:** 1894–1904.
14. Zhou, B., Q. Ma, S. Rajagopal, *et al.* 2008. Epicardial progenitors contribute to the cardiomyocyte lineage in the developing heart. *Nature* **454:** 109–113.
15. Smart, N., S. Bollini, K.N. Dube, *et al.* 2011. De novo cardiomyocytes from within the activated adult heart after injury. *Nature* **474:** 640–644.
16. Smart, N. & P.R. Riley. 2009. Derivation of epicardium-derived progenitor cells (EPDCs) from adult epicardium. *Curr. Protoc. Stem Cell Biol.* **Chapter 2:** Unit2C.
17. Moretti, A., L. Caron, A. Nakano, *et al.* 2006. Multipotent embryonic isl1+ progenitor cells lead to cardiac, smooth muscle, and endothelial cell diversification. *Cell* **127:** 1151–1165.
18. Zhou, B., L.B. Honor, Q. Ma, *et al.* 2011. Thymosin beta 4 treatment after myocardial infarction does not reprogram epicardial cells into cardiomyocytes. *J. Mol. Cell Cardiol.* **52:** 43–47.
19. Bock-Marquette, I., A. Saxena, M.D. White, *et al.* 2004. Thymosin β4 activates integrin-linked kinase and promotes cardiac cell migration, survival and cardiac repair. *Nature* **432:** 466–472.
20. Bock-Marquette, I., S. Shrivastava, G.C. Pipes, *et al.* 2009. Thymosin beta4 mediated PKC activation is essential to initiate the embryonic coronary developmental program and epicardial progenitor cell activation in adult mice in vivo. *J. Mol. Cell Cardiol.* **46:** 728–738.
21. Sutton, M.G. & N. Sharpe. 2000. Left ventricular remodeling after myocardial infarction: pathophysiology and therapy. *Circulation* **101:** 2981–2988.
22. Frangogiannis, N.G. 2006. Targeting the inflammatory response in healing myocardial infarcts. *Curr. Med. Chem.* **13:** 1877–1893.
23. Frohm, M., H. Gunne, A.C. Bergman, *et al.* 1996. Biochemical and antibacterial analysis of human wound and blister fluid. *Eur. J. Biochem.* **237:** 86–92.
24. Sosne, G., C. Chan, K. Thai, *et al.* 2001. Thymosin beta 4 promotes corneal wound healing and modulates inflammatory mediators in vivo. *Exp. Eye Res.* **72:** 605–608.
25. Girardi, M., M.A. Sherling, R.B. Filler, *et al.* 2003. Anti-inflammatory effects in the skin of thymosin-beta4 splice-variants. *Immunology* **109:** 1–7.
26. Sosne, G., E.A. Szliter, R. Barrett, *et al.* 2002. Thymosin beta 4 promotes corneal wound healing and decreases inflammation in vivo following alkali injury. *Exp. Eye Res.* **74:** 293–299.
27. Badamchian, M., M.O. Fagarasan, R.L. Danner, *et al.* 2003. Thymosin beta(4) reduces lethality and down-regulates inflammatory mediators in endotoxin-induced septic shock. *Int. Immunopharmacol.* **3:** 1225–1233.
28. Zhang, J., Z.G. Zhang, D. Morris, *et al.* 2009. Neurological functional recovery after thymosin beta4 treatment in mice with experimental auto encephalomyelitis. *Neuroscience* **164:** 1887–1893.

29. Sosne, G., P. Qiu, P.L. Christopherson & M.K. Wheater. 2007. Thymosin beta 4 suppression of corneal NFkappaB: a potential anti-inflammatory pathway. *Exp. Eye Res.* **84:** 663–669.

30. Sopko, N., Y. Qin, A. Finan, *et al.* 2011. Significance of thymosin beta4 and implication of PINCH-1-ILK-alpha-parvin (PIP) complex in human dilated cardiomyopathy. *PLoS One* **6:** e20184.

31. Qiu, P., M.K. Wheater, Y. Qiu & G. Sosne. 2011. Thymosin beta4 inhibits TNF-alpha-induced NF-kappaB activation, IL-8 expression, and the sensitizing effects by its partners PINCH-1 and ILK. *FASEB J.* **25:** 1815–1826.

32. Peng, H., O.A. Carretero, L. Raij, *et al.* 2001. Antifibrotic effects of N-acetyl-seryl-aspartyl-Lysyl-proline on the heart and kidney in aldosterone-salt hypertensive rats. *Hypertension* **37:** 794–800.

33. Qian, L., Y. Huang, C.I. Spencer, *et al.* 2012. In vivo reprogramming of murine cardiac fibroblasts into induced cardiomyocytes. *Nature* **485:** 593–598.

Ann. N.Y. Acad. Sci. ISSN 0077-8923

ANNALS OF THE NEW YORK ACADEMY OF SCIENCES

Issue: *Thymosins in Health and Disease*

Molecular and cellular mechanisms of thymosin β4–mediated cardioprotection

Rabea Hinkel,[1] Teresa Trenkwalder,[1] and Christian Kupatt[1,2]

[1]Medizinische Klinik und Poliklinik I, Klinikum Großhadern, Ludwig Maximilians University, Munich, Germany. [2]Munich Heart Alliance, Munich, Germany

Address for correspondence: Rabea Hinkel, DVM, Medizinische Klinik und Poliklinik I, Klinikum Großhadern, Ludwig-Maximilians-University, Marchioninistr 15, 81377 Munich, Germany. rabea.hinkel@med.uni-muenchen.de

Coronary heart disease is still the leading cause of death in industrialized nations. Reduction of infarct size after acute myocardial infarction and, in addition, improvement of myocardial function and perfusion in acute and chronic myocardial ischemia would enhance cardiac survival. Thymosin β4, a 43-amino acid water-soluble peptide with pleiotropic abilities seems to be a promising candidate for the treatment of ischemic heart disease. During cardiac development, thymosin β4 is essential for vascularization of the myocardium, by targeting all three parts of vessel development, that is, vasculogenesis, angiogenesis, and arteriogenesis. In the adult, thymosin β4 is capable of inducing angiogenesis via activation of survival kinases in an actin-dependent and -independent manner. In addition, thymosin β4 has anti-inflammatory properties by reducing NF-κB p65 activation. These protective effects are further enhanced through increased myocyte and endothelial cell survival accompanied by differentiation of epicardial progenitor cells.

Keywords: thymosin β4; ischemia/reperfusion injury; inflammation; angiogenesis; cardioprotection

Introduction

Coronary artery disease (CAD) and its consequence—cardiac heart failure—remain the main cause of death in industrial nations.[1] A characteristic manifestation of CAD is an acute myocardial infarction, which is caused by acute and ongoing occlusion of the coronary artery. On the other hand, chronic ischemic cardiomyopathy is caused by a gradual narrowing and eventual occlusion of one or more coronary arteries.

Treatment of choice for acute myocardial infarction is the revascularization of the stenosed or occluded coronary artery. The revascularization is performed either via percutaneous transluminal coronary angioplasty or bypass surgery and reduces cardiac related mortality.[1–3] However, the revascularization itself leads to an additional loss of myocytes termed *ischemia-reperfusion injury*,[4] which counteracts the beneficial effect of myocardial reperfusion.[5] Ischemia-reperfusion injury is characterized by myocardial stunning, increased reactive oxygen species (ROS) levels, enhanced inflammation, and cell apoptosis.[5] In the experimental setting of coronary ischemia-reperfusion, treatment of ischemia-reperfusion injury is capable of further reducing the infarct size, in addition to the protective effect of reperfusion itself.[5]

In contrast to the pathophysiological mechanisms of ischemia-reperfusion injury, chronic ischemic cardiomyopathy is characterized by a left ventricular impairment, a chronic malperfusion of the myocardium inducing rarification of the capillary density and hibernating myocardium.[6] Hibernating myocardium is distinguished by viable myocytes that, however, do not contribute to the pump function of the heart.[6] Since these myocytes remain viable over a prolonged period of time, reestablishing perfusion might reactivate these myocytes and thereby improve myocardial function. A growing number of patients suffering from hibernating myocardium are not suitable for revascularization

doi: 10.1111/j.1749-6632.2012.06693.x

Ann. N.Y. Acad. Sci. 1269 (2012) 102–109 © 2012 New York Academy of Sciences.

(interventional as well as surgical). For these so called "no option" patients, therapeutic neovascularization via proangiogenic gene therapy seems to be a promising option.

In order to provide an adequate therapy of ischemia-reperfusion injury as well as chronic ischemic cardiomyopathy, a potential therapeutic factor has to fulfill some characteristics, such as anti-inflammatory, antiapoptotic, and proangiogenic properties. All of these features have been shown for thymosin β4, a 43-amino acid peptide, which might fulfill the necessities for cardioprotection in either an acute or chronic ischemic coronary pathological entity.

Thymosin β4 and cardiac development

During embryonic development in mice, thymosin β4 is upregulated in regions of blood vessel formation as well as in the endocardial cushions.[7] Embryonic endothelial progenitor cells, isolated from the regions of vasculogenesis at day E7.5, have significantly higher levels of thymosin β4 than adult endothelial cells.[8,9] Furthermore, Bock-Marquette *et al.* found thymosin β4 to be expressed in the left ventricle, the outer curvature of the right ventricle, and the cardiac outflow tract in mice at day E11.5.[10] During embryonic development (days) E9.5–12.5 thymosin β4 was expressed in the ventricular septum and regions of high proliferation, pointing to an essential role of thymosin β4 in cardiac development in general.[10]

Vessel development in embryos includes vasculogenesis, angiogenesis, as well as arteriogenesis, together leading to a mature and stable vessel network.[11] Besides endothelial cells, smooth muscle cells are of utmost important, since they stabilize the vessels during arteriogenesis.[11] Smart *et al.* demonstrated that thymosin β4 is involved in all three steps of vessel development in mice embryos (Fig. 1).[12] Mice with a thymosin β4 knockdown displayed a reduced microvessel density in the myocardium (impaired vasculogenesis), malformation of coronary vessels (angiogenesis) and coverage of smooth muscle cells (arteriogenesis).[12] In contrast, thymosin β4 knockout mice, either global or cardiac specific (Nkx 2.5-Cre and α MHC-Cre), showed no impaired cardiac development or increased embryonic lethality.[13] Furthermore, adult cardiac function, coronary vascular density and volume in thymosin β4 knockout mice were not altered com-

pared to wild type mice.[13] These different findings between knockout mice and shRNA knockdown mice might be due to different compensatory mechanisms or off-target effects in the two different strategies. Furthermore, these findings are obtained under physiological conditions and the influence of thymosin β4 knockout in acute or chronic ischemia needs to be examined in further studies.

The effects in thymosin knockdown mice are triggered by an impaired migration of epicardium-derived cells, which have the capability to differentiate into either endothelial cells by VEGF and bFGF or smooth muscle cells stimulated by PDGF and TGF-β.[14] Interestingly, this vascular differentiation requires thymosin β4 in the myocyte compartment and not in the endothelial cells or the epicardium, which seems to trigger the epicardial cell migration into the inner layers of the myocardium, pointing out a paracrine effect of thymosin β4 on vessel development.[12] Paracrine myocyte derived signaling for vessel development in the heart was also described for VEGF and FOG-2, but in contrast to thymosin β4, only vasculogenesis was affected, whereas arteriogenesis and maturation of the vessels were found unaltered.[15,16] Beside the paracrine activation of epicardium-derived cells, thymosin β4 induces embryonic cardiomyocyte and endothelial cell migration, improves cell survival, and increases the beating frequency of the cardiomyocytes.[10,17]

Thymosin β4 and angiogenesis

Grant *et al.* first discovered the relation between thymosin β4 and angiogenesis in 1995.[18] They found, that HUVECs plated on matrigel rapidly induced thymosin β4 and that transfection increased tube formation *in vitro*[18] (Fig. 1). Interestingly, some of the β-thymosins promote angiogenesis, where as other inhibit angiogenesis. In a chicken chorioallantois membrane model, thymosin β4, and thymosin β15 promote angiogenesis in contrast to thymosin β9 and thymosin β10 that inhibit angiogenesis.[19–22] Furthermore, thymosin β4 is a chemoattractant for endothelial cell migration *in vitro* and *in vivo* in a matrigel plaque assay,[23] and induces sprouting of aortic rings.[24] Since migration is an essential functional element of angiogenesis and thymosin β4 is an important G-actin–sequestering peptide responsible for rapid actin reorganization in the cell an

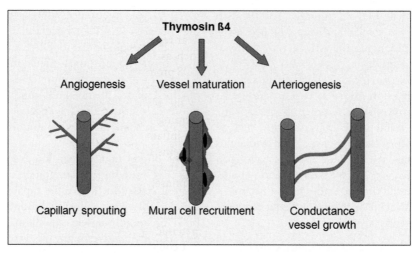

Figure 1. Thymosin β4 is essential for three mechanisms of vessel formation and outgrowth in the developing heart. It stimulates endothelial cells to form new sprouts resulting in angiogenesis. In parallel, it stabilizes cells by recruiting mural cells, a process being of utmost importance for a long-lasting vascular network. Furthermore, thymosin β4 induces conductance vessel growth, representing the elementary process for sufficient perfusion.

important question is, if the proangiogenic potential of thymosin β4 is actin dependent. Philp *et al.* investigated the role of actin binding in thymosin β4-mediated angiogenesis.[25] *In vitro* studies and chick aortic arch assays displayed an actin binding motif-dependent role of thymosin β4 in angiogenesis, whereas synthetic peptides lacking the actin binding motif were inactive.[25] Even though the actin-binding motif is well conserved within the β thymosin family[26] and is described to be responsible for the proangiogenic abilities of thymosin β4 and 15, there must be other proangiogenic effects independent of the actin binding site, since thymosin β9 and β10 are antiangiogenic but have the same actin binding motif as the other two proangiogenic β thymosins. In summary, besides the actin-dependent angiogenic effect, there might be an actin-independent proangiogenic effect of thymosin β4.[27] One possible proangiogenic actin-independent pathway of thymosin β4 might be the induction of VEGF, a well-known angiogenic factor, via Hif-1α stabilization.[28] Another actin-independent proangiogenic mechanism thymosin, β4 is described by Freeman *et al.*[29] They demonstrated that thymosin β4 acts as a regulator of the purinergic signaling, by increasing the cell surface ATP levels via ATP synthase after extracellular application.[29] Futhermore, intracellular thymosin β4 forms a functional complex

with PINCH and ILK, thereby activating the protein kinase B, a survival kinase also promoting angiogenesis.[10] Thymosin β4 enhanced endothelial progenitor cell migration leads to a concentration-dependent Akt, eNOS, and Erk1/2 phosphorylation.[30] Inhibition of Akt and eNOS via unspecific inhibitors abolished thymosin β4-induced cell migration, whereas inhibition of the Map-kinases had no effect.[30] Since the study of Asahara *et al.*, it is well known that endothelial progenitor cells are capable of inducing angiogenesis.[31] Embryonic endothelial progenitor cells induce angiogenesis *in vitro* and *in vivo* in a chronic ischemia model.[9] These protective effects of eEPCs are thymosin β4 dependent, since downregulation of thymosin β4 via shRNA in these cells abolishes the proangiogenic effect.[32]

With regard to arteriogenesis, a study of Bicer *et al.* demonstrated that the collateralization of patients with chronic ischemic heart disease correlates with their systemic thymosin β4 levels in the serum.[33] In patients with a well-developed collateralization of thymosin β4 concentration in the serum was doubled compared to the poorly developed collateral group.[33] All other baseline characteristics and risk factors, such as diabetes or hyperlipidemia, did not differ between the two groups.[33] These results were confirmed by Lv *et al.* in a larger patient cohort.[34] Taken together, thymosin β4 has proangiogenic and proarteriogenic abilities, which seem to be actin-binding dependent as well as actin-binding independent. Activation of proangiogenic survival kinases as well as improvement of migrational potential of angiogenic cells is responsible for this profound angiogenic effect. Additionally, the angiogenic potential of endothelial progenitor cells depends on thymosin β4.

Thymosin β4 and inflammation

Inflammation plays a major role in acute and chronic myocardial ischemia. During acute myocardial infarction, leukocyte influx is triggered by the ischemia-reperfusion injury.[35] In addition, in atherosclerotic lesions increasing numbers of proinflammatory cells destabilize the plaque, leading to a plaque rupture, and acute occlusion of the coronary artery.[36] In the chronic disease progression, macrophages perpetuate vascular inflammation and stenosis progression, therefore, in respect to a potential therapeutic use of thymosin β4; its influence on inflammation is of great interest.

In models of corneal wound healing, thymosin β4 not only promoted reepithelialization, but also reduced inflammation via downregulation of proinflammatory cytokines (IL-1β, IL-18) and chemokines (MIP 1α, 1β, 2).[37,38] Even in an advanced cornea injury model of alkali injury, thymosin β4 is capable of reducing the inflammatory answer (cytokine levels and PMNs).[38] After TNF-α stimulation the NF-κB p65 activation and nuclear translocation is significantly reduced via thymosin β4.[39] Ping *et al.* demonstrated that thymosin β4 inhibits TNF-α induced NF-κB activation and IL8 induction via the PINCH-ILK complex, known as focal adhesion complex proteins contributing to TNF-α signaling.[40,41] The use of an actin-binding–deficient thymosin β4 mutant revealed similar results. Therefore, the blockage NF-κB activation and translocation by thymosin β4 is an actin-independent mechanism.[40] In addition, Young *et al.* demonstrated that thymosin β4 sulfoxide is released from glucocorticoid stimulated macrophages, pointing to an extracellular bioactivity of thymosin β4.[42] This oxidation at Met-6 of thymosin β4, a residue not essential for the actin-binding domain, attenuates the neutrophil-associated inflammatory response *in vitro* and *in vivo*.[42] Furthermore, this oxidation and reduction of

the methionine residue could be responsible for cellular regulatory mechanisms during inflammation, such as increased F-actin formation and reduced ROS production in response to oxidative stress.[42] In contrast, it is well known that G-actin is released to the blood after infection or from dying cells, forming F-actin filaments and thereby promoting the inflammatory reaction.[43] G-actin–sequestering proteins, such as thymosin β4, act as actin-scavenger and thereby counteract toxicity and infectious actin response, pointing to an actin-dependent protective effect.[44] In sepsis, levels of cytokines and inflammatory molecules are increased, whereas systemic thymosin β4 levels are reduced. This effect could be simulated in healthy volunteers after LPS injection.[44] In a model of LPS-induced sepsis in mice, thymosin β4 application after sepsis induction significantly reduced cytokine levels and enhanced survival.[44] Activated platelets release thymosin β4, which is attached to factor VIII, fibrin, and collagen and thereby leading to high concentration of thymosin β4 in the thrombus, promoting wound healing.[45,46] After acute myocardial infarction in pigs, regional thymosin β4 application significantly reduced the leukocyte influx in the ischemic area. This attenuation of the local inflammation reduces the infarct size and improves myocardial function after 24 hours (Fig. 2).[47]

Thymosin β4 and cardioprotection

Since it is known that thymosin β4 plays an important role in cardiac development, the question arose, whether it has also cardioprotective properties in the adult organism. Bock-Marquette *et al.* demonstrated that a systemic application of thymosin β4 peptide in mice undergoing left anterior descending artery (LAD) ligation, significantly reduced infarct size and improved myocardial function.[10] These findings are based on the interaction with PINCH and ILK, thereby phosphorylating Akt and enhancing cardiomyocyte survival.[10] In a preclinical model of ischemia/reperfusion injury, local application of thymosin β4 peptide significantly reduced the infarct size an enhanced myocardial function. These effects were based on a reduced endothelial cell and myocyte apoptosis after hypoxia/reoxygenation as well as on reduced leukocyte recruitment *in vitro* and *in vivo*[32,47] (Fig. 2).

A potential mechanism for the cardioprotective effect of thymosin β4 is the outgrowth of adult

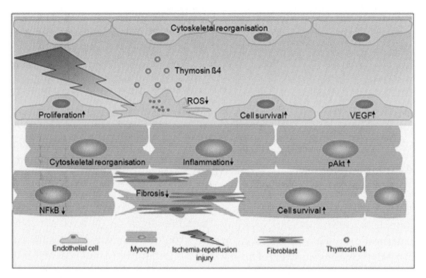

Figure 2. Ischemia-reperfusion injury is characterized by acute inflammatory status, as well as ROS production leading to myocyte and endothelial cell apoptosis. Without therapy, this process results in irreversible loss of myocytes, myocardial fibrosis resulting in heart failure. Thymosin β4 is a promising candidate for ischemia-reperfusion injury, as it targets these pathophysiological mechanisms. It reduces NF-κB activation and ROS production, thereby reducing inflammation. Furthermore endothelial cell survival, migration, and proliferation are enhanced via thymosin β4. In myocytes, Akt activation significantly enhances cell survival.

epicardial cells after thymosin β4 treatment.[12,14] Thymosin β4 restores pluripotency in these outgrowing cells and triggers differentiation into fibroblasts, smooth muscle cells and endothelial cells.[12] A cleavage product of thymosin β4, AcSDKP, is not capable of restoring all thymosin β4 functions. In thymosin β4 mutant mice, AcSDKP application is not capable of rescuing the phenotype, but still has cardioprotective effects in the adult.[12] AcSDKP induces endothelial cell differentiation out of the epicardial progenitor cells, but fails to induce smooth-muscle cells and fibroblasts.[12] Besides the *ex vivo* outgrowth, priming with thymosin β4 induces the reexpression of Wilm's tumor 1 (Wt1) in epicardial progenitor cells. The induction of Wt1 leads to a mobilization of these cells and a differentiation to *de novo* cardiomyocytes.[48] Analysis of thymosin β4 effects on the whole heart displayed an organ-wide activation of the embryonic coronary development program in the ischemic heart, such as upregulation of VEGF, VEGF-R2, TGF-β, FGF17, as well as β-catenin and increased Tbx18 and Wt-1 cells in the myocardium.[17] This induces an activation of the epicardial progenitor cells and leading to an enhanced cardiac regeneration in a PKC-dependent manner, since inhibition of PKC

abolished these effects.[17] Applying thymosin β4 in a more clinical setting, after acute myocardial infarction, Zhou *et al.* found a reduced infarct size, cardiomyocyte apoptosis, fibrosis, as well as an enhanced capillary density.[49] Utilizing the same Wt1-based reporer mice as in the study of Smart *et al.*, they confirmed an increased thickness of the epicardium and a differentiation of the epicardial cells to fibroblasts and smooth muscle cells, but did not observe novo cardiomyocytes and coronary endothelial cells formation.[49] Whether these differences in reprogramming of the epicardial cells toward cardiomyocytes are based only on the different treatment timepoints, and thereby a different physiological response of the epicardium pre- and post-MI to the thymosin β4, needs to be evaluated in further studies.[50]

Myocardial remodeling and fibrosis after acute myocardial infarction or chronic myocardial ischemia are the key mechanisms for the development of heart failure. Since thymosin β4 is capable of inducing fibroblast differentiation from epicardial progenitor cells, its effect on fibroblasts in ischemia is of particular interest. Treatment of fibroblasts with thymosin β4 before adding low concentrations of H_2O_2 stimulating ischemia-reoxygenation significantly reduced the reactive oxygen species (ROS) *in vitro*.[51] It increases the expression of antioxidative genes and meanwhile reduces the expression of profibrotic genes in fibroblasts after ischemia.[51] *In vivo* thymosin β4 enhances the focal adhesion complex PIP (PINCH-ILK-α parvin), which plays a critical role in the development and function of the heart.[52] This focal adhesion complex activates Akt and suppressing NF-κB and collagen (fibrosis) (Fig. 2).[52]

In summary, we and others have demonstrated that thymosin β4 has cardioprotective abilities after ischemia and ischemia-reperfusion injury. These cardioprotective effects are enabled by a reduced inflammatory reaction and an increased activation of survival kinases, improving endothelial cell, and cardiomyocyte survival. In a phase 1 study, safety for intravenous application of thymosin β4 (single dosage 1260 mg) was demonstrated. No evidence for toxicity or side effects were observed.[53] These results suggest that thymosin β4 is a promising peptide for ischemia-reperfuison injury treatment, since it targets the crucial pathophysiological mechanisms (Fig. 2) (for review, see Ref. 54). In terms of chronic myocardial ischemia, thymosin β4 reduces fibrosis and induces vessel growth. Thymosin β4-mediated vessel formation is based on angiogenesis, vessel maturation, and arteriogenesis in the developing heart. These effects of thymosin β4 seem to be actin dependent, since using actin-deficient thymosin β4 mutants only restored part of the proangiogenic effects. In therapeutic neovascularization, vessel maturation and arteriogenesis seem to be important for long-lasting efficient restoration of perfusion, in addition to angiogenesis (Fig. 1).

In conclusion, thymosin β4 seems to be a promising therapeutic approach for the treatment of acute ischemia-reperfusion injury as well as for chronic ischemic cardiomyopathy.

Acknowledgments

C.K. is supported by the Deutsche Forschungsgemeinschaft by the German Ministry for Research and Education and by the Else-Kröner-Fresenius Stiftung. R.H. is supported by the FöFoLe and the Friedrich-Baur-Stiftung.

Conflicts of interest

The authors declare no conflicts of interest.

References

1. Lloyd-Jones, D., R.J. Adams, T.M. Brown, *et al.* 2010. Heart disease and stroke statistics—2010 update: a report from the American Heart Association. *Circulation* **121:** e246–e215.
2. Suero, J.A., S.P. Marso, P.G. Jones, *et al.* 2001. Procedural outcomes and long-term survival among patients undergoing percutaneous coronary intervention of a chronic total occlusion in native coronary arteries: a 20-year experience. *J. Am. Coll. Cardiol.* **38:** 409–414.
3. White, H.D. & D.P. Chew. 2008. Acute myocardial infarction. *Lancet* **372:** 570–584.
4. Yellon, D.M. & G.F. Baxter. 1999. Reperfusion injury revisited: is there a role for growth factor signaling in limiting lethal reperfusion injury? *Trends Cardiovasc. Med.* **9:** 245–249.
5. Yellon, D.M. & D.J. Hausenloy. 2007. Myocardial reperfusion injury. *N. Engl. J. Med.* **357:** 1121–1135.
6. Elsasser, A., E. Decker, S. Kostin, *et al.* 2000. A self-perpetuating vicious cycle of tissue damage in human hibernating myocardium. *Mol. Cell Biochem.* **213:** 17–28.
7. Gomez-Marquez, J., A.F. Franco del, P. Carpintero & R. Anadon. 1996. High levels of mouse thymosin beta4 mRNA in differentiating P19 embryonic cells and during development of cardiovascular tissues. *Biochim. Biophys. Acta* **1306:** 187–193.
8. Hatzopoulos, A.K., J. Folkman, E. Vasile, *et al.* 1998. Isolation and characterization of endothelial progenitor cells from mouse embryos. *Development* **125:** 1457–1468.
9. Kupatt, C., J. Horstkotte, G.A. Vlastos, *et al.* 2005. Embryonic endothelial progenitor cells expressing a broad range of proangiogenic and remodeling factors enhance vascularization and tissue recovery in acute and chronic ischemia. *FASEB J.* **19:** 1576–1578.
10. Bock-Marquette, I., A. Saxena, M.D. White, *et al.* 2004. Thymosin beta4 activates integrin-linked kinase and promotes cardiac cell migration, survival and cardiac repair. *Nature* **432:** 466–472.
11. Carmeliet, P. 2000. Mechanisms of angiogenesis and arteriogenesis. *Nat. Med.* **6:** 389–395.
12. Smart, N., C.A. Risebro, A.A.D. Melville, *et al.* 2007. Thymosin beta4 induces adult epicardial progenitor mobilization and neovascularization. *Nature* **445:** 177–182.
13. Banerjee, I., J. Zhang, T. Moore-Morris, *et al.* 2012. Thymosin beta 4 is dispensable for murine cardiac development and function. *Circ. Res.* **110:** 456–464.
14. Smart, N., C.A. Risebro, A.A. Melville, *et al.* 2007. Thymosin beta-4 is essential for coronary vessel development and promotes neovascularization via adult epicardium. *Ann. N.Y. Acad. Sci.* **1112:** 171–188.
15. Tevosian, S.G., A.E. Deconinck, *et al.* 2000. FOG-2, a cofactor for GATA transcription factors, is essential for heart morphogenesis and development of coronary vessels from epicardium. *Cell* **101:** 729–739.
16. Giordano, F.J., H.P. Gerber, S.P. Williams, *et al.* 2001. A cardiac myocyte vascular endothelial growth factor paracrine pathway is required to maintain cardiac function. *Proc. Natl. Acad. Sci. USA* **98:** 5780–5785.
17. Bock-Marquette, I., S. Shrivastava, G.C. Pipes, *et al.* 2009. Thymosin beta4 mediated PKC activation is essential to

initiate the embryonic coronary developmental program and epicardial progenitor cell activation in adult mice *in vivo. J. Mol. Cell Cardiol.* **46:** 728–738.

18. Grant, D.S., J.L. Kinsella, M.C. Kibbey, *et al.* 1995. Matrigel induces thymosin beta 4 gene in differentiating endothelial cells. *J. Cell Sci.* **108:** 3685–3694.

19. Koutrafouri, V., L. Leondiadis, K. Avgoustakis, *et al.* 2001. Effect of thymosin peptides on the chick chorioallantoic membrane angiogenesis model. *Biochim. Biophys. Acta.* **1568:** 60–66.

20. Koutrafouri, V., L. Leondiadis, N. Ferderigos, *et al.* 2003. Synthesis and angiogenetic activity in the chick chorioallantoic membrane model of thymosin beta-15. *Peptides* **24:** 107–115.

21. Lee, S.H., M.J. Son, S.H. Oh, *et al.* 2005. Thymosin {beta}(10) inhibits angiogenesis and tumor growth by interfering with Ras function. *Cancer Res.* **65:** 137–148.

22. Mu, H., R. Ohashi, H. Yang, *et al.* 2006. Thymosin beta10 inhibits cell migration and capillary-like tube formation of human coronary artery endothelial cells. *Cell Motil. Cytoskeleton* **63:** 222–230.

23. Malinda, K.M., A.L. Goldstein & H.K. Kleinman. 1997. Thymosin beta 4 stimulates directional migration of human umbilical vein endothelial cells. *FASEB J.* **11:** 474–481.

24. Grant, D.S., W. Rose, C. Yaen, *et al.* 1999. Thymosin beta4 enhances endothelial cell differentiation and angiogenesis. *Angiogenesis* **3:** 125–135.

25. Philp, D., T. Huff, Y.S. Gho, *et al.* 2003. The actin binding site on thymosin beta4 promotes angiogenesis. *FASEB J.* **17:** 2103–2105.

26. Huff, T., C.S. Muller, A.M. Otto, *et al.* 2001. Beta-thymosins, small acidic peptides with multiple functions. *Int. J. Biochem. Cell Biol.* **33:** 205–220.

27. Smart, N., A. Rossdeutsch & P.R. Riley. 2007. Thymosin beta4 and angiogenesis: modes of action and therapeutic potential. *Angiogenesis* **10:** 229–241.

28. Jo, J.O., S.R. Kim, M.K. Bae, *et al.* 2010. Thymosin beta4 induces the expression of vascular endothelial growth factor (VEGF) in a hypoxia-inducible factor (HIF)-1alpha-dependent manner. *Biochim. Biophys. Acta* **1803:** 1244–1251.

29. Freeman, K.W., B.R. Bowman & B.R. Zetter. 2011. Regenerative protein thymosin beta-4 is a novel regulator of purinergic signaling. *FASEB J.* **25:** 907–915.

30. Qiu, F.Y., X.X. Song, H. Zheng, *et al.* 2009. Thymosin beta4 induces endothelial progenitor cell migration via PI3K/Akt/eNOS signal transduction pathway. *J. Cardiovasc. Pharmacol.* **53:** 209–214.

31. Asahara, T., T. Murohara, A. Sullivan, *et al.* 1997. Isolation of putative progenitor endothelial cells for angiogenesis. *Science* **275:** 964–967.

32. Hinkel, R., I. Bock-Marquette, A.K. Hatzopoulos & C. Kupatt. 2010. Thymosin beta4: a key factor for protective effects of eEPCs in acute and chronic ischemia. *Ann. N.Y. Acad. Sci.* **1194:** 105–111.

33. Bicer, A., O. Karakurt, R. Akdemir, *et al.* 2011. Thymosin beta 4 is associated with collateral development in coronary artery disease. *Scand. J. Clin. Lab Invest.* **71:** 625–630.

34. Lv, S., G. Cheng, Y. Xu, *et al.* 2011. Relationship between serum thymosin beta4 levels and coronary collateral development. *Coron. Artery Dis.* **22:** 401–404.

35. Vinten-Johansen, J. 2004. Involvement of neutrophils in the pathogenesis of lethal myocardial reperfusion injury. *Cardiovasc. Res.* **61:** 481–497.

36. van, R.N., I. Hoefer, I. Buschmann, *et al.* 2003. Effects of local MCP-1 protein therapy on the development of the collateral circulation and atherosclerosis in Watanabe hyperlipidemic rabbits. *Cardiovasc. Res.* **57:** 178–185.

37. Sosne, G., C.C. Chan, K. Thai, *et al.* 2001. Thymosin beta 4 promotes corneal wound healing and modulates inflammatory mediators *in vivo. Exp. Eye Res.* **72:** 605–608.

38. Sosne, G., E.A. Szliter, R. Barrett, *et al.* 2002. Thymosin beta 4 promotes corneal wound healing and decreases inflammation *in vivo* following alkali injury. *Exp. Eye Res.* **74:** 293–299.

39. Sosne, G., P. Qiu, P.L. Christopherson & M.K. Wheater. 2007. Thymosin beta 4 suppression of corneal NFkappaB: a potential anti-inflammatory pathway. *Exp. Eye Res.* **84:** 663–669.

40. Qiu, P., M.K. Wheater, Y. Qiu & G. Sosne. 2011. Thymosin beta4 inhibits TNF-alpha-induced NF-kappaB activation, IL-8 expression, and the sensitizing effects by its partners PINCH-1 and ILK. *FASEB J.* **25:** 1815–1826.

41. Babakov, V.N., O.A. Petukhova, L.V. Turoverova, *et al.* 2008. RelA/NF-kappaB transcription factor associates with alpha-actinin-4. *Exp. Cell Res.* **314:** 1030–1038.

42. Young, J.D., A.J. Lawrence, A.G. MacLean, *et al.* 1999. Thymosin beta 4 sulfoxide is an anti-inflammatory agent generated by monocytes in the presence of glucocorticoids. *Nat. Med.* **5:** 1424–1427.

43. Lee, J.T. 1992. Antibiotic prophylaxis and surgical-wound infections. *N. Engl. J. Med.* **327:** 205.

44. Badamchian, M., M.O. Fagarasan, R.L. Danner, *et al.* 2003. Thymosin beta(4) reduces lethality and down-regulates inflammatory mediators in endotoxin-induced septic shock. *Int. Immunopharmacol.* **3:** 1225–1233.

45. Huff, T., A.M. Otto, C.S. Muller, *et al.* 2002. Thymosin beta4 is released from human blood platelets and attached by factor XIIIa (transglutaminase) to fibrin and collagen. *FASEB J.* **16:** 691–696.

46. Mannherz, H.G. & E. Hannappel. 2009. The beta-thymosins: intracellular and extracellular activities of a versatile actin binding protein family. *Cell Motil. Cytoskeleton* **66:** 839–851.

47. Hinkel, R., C. El-Aouni, T. Olson, *et al.* 2008. Thymosin beta 4 is an essential paracrine factor of embryonic endothelial progenitor cell-mediated cardioprotection. *Circulation* **117:** 2232–2240.

48. Smart, N., S. Bollini, K.N. Dube, *et al.* 2011. De novo cardiomyocytes from within the activated adult heart after injury. *Nature* **474:** 640–644.

49. Zhou, B., L.B. Honor, Q. Ma, *et al.* 2012. Thymosin beta 4 treatment after myocardial infarction does not reprogram epicardial cells into cardiomyocytes. *J. Mol. Cell Cardiol.* **52:** 43–47.

50. Kispert, A. 2012. No muscle for a damaged heart: thymosin beta 4 treatment after myocardial infarction does not induce myocardial differentiation of epicardial cells. *J. Mol. Cell Cardiol.* **52:** 10–12.

51. Kumar, S. & S. Gupta. 2011. Thymosin beta 4 prevents oxidative stress by targeting antioxidant and anti-apoptotic genes in cardiac fibroblasts. *PLoS One* **6:** e26912.

52. Sopko, N., Y. Qin, A. Finan, *et al.* 2011. Significance of thymosin beta4 and implication of PINCH-1-ILK-alpha-parvin (PIP) complex in human dilated cardiomyopathy. *PLoS One* **6:** e20184.

53. Ruff, D., D. Crockford, G. Girardi & Y. Zhang. 2010. A randomized, placebo-controlled, single and multiple dose study of intravenous thymosin beta4 in healthy volunteers. *Ann. N.Y. Acad. Sci.* **1194:** 223–229.

54. Dube, K.N., S. Bollini, N. Smart & P.R. Riley. 2012. Thymosin beta4 protein therapy for cardiac repair. *Curr. Pharm. Des* **18:** 799–806.

Ann. N.Y. Acad. Sci. ISSN 0077-8923

ANNALS OF THE NEW YORK ACADEMY OF SCIENCES
Issue: *Thymosins in Health and Disease*

Treatment of neurological injury with thymosin β4

Daniel C. Morris,[1] Zheng G. Zhang,[2] Jing Zhang,[2] Ye Xiong,[3] Li Zhang,[2] and Michael Chopp[2,4]

[1]Department of Emergency Medicine, [2]Department of Neurology, [3]Department of Neurosurgery, Henry Ford Health Sciences Center, Detroit, Michigan. [4]Department of Physics, Oakland University, Rochester, Michigan

Address for correspondence: Daniel C. Morris, M.D., Henry Ford Hospital, 2799 West Grand Blvd, Department of Emergency Medicine, CFP-2, Detroit, MI 48202. Dmorris4@hfhs.org

Neurorestorative therapy targets multiple types of parenchymal cells in the intact tissue of injured brain tissue to increase neurogenesis, angiogenesis, oligodendrogenesis, and axonal remodeling during recovery from neurological injury. In our laboratory, we tested thymosin β4 (Tβ4) as a neurorestorative agent to treat models of neurological injury. This review discusses our results demonstrating that Tβ4 improves neurological functional outcome in a rat model of embolic stroke, a mouse model of multiple sclerosis, and a rat model of traumatic brain injury. Tβ4 is a pleiotropic peptide exhibiting many actions in several different types of tissues. One mechanism associated with improvement of neurological improvement from Tβ4 treatment is oligodendrogenesis involving the differentiation of oligodendrocyte progenitor cells to mature myelin-secreting oligodendrocytes. Moreover, our preclinical data provide a basis for movement of Tβ4 into clinical trials for treatment of these devastating neurological diseases and injuries.

Keywords: thymosin β4; stroke; multiple sclerosis; traumatic brain injury; rat

Treatment of neurological injury remains an elusive goal in health care. Despite billions of dollars invested in basic neuroscience and clinical trials, meaningful treatment of many neurological diseases remains a difficult task. As the population ages, certain neurological illnesses, such as stroke and dementia, will increasingly cause severe neurologic disability.[1,2] At an individual level, the overall quality of life of those stricken with these diseases as well as those who care for them is severely affected both financially and emotionally. At the national level, the projected rate of health-care spending will be severely affected unless meaningful treatments are discovered and implemented.[3]

Neurorestorative therapy is a concept that is just beginning to be recognized at both the basic science and clinical levels. Neurorestorative therapy treats the intact tissue and not the lesion to invoke a repair of damaged tissues from any type of neurological injury.[4] Neurorestorative therapy acts on intact parenchymal cells, specifically neuroprogenitor cells (adult neural stem cells), oligodendrocyte progenitor cells (OPCs), astroglial cells, and cerebral endothelial cells, to promote neurogenesis, oligodendrogenesis, axonal sprouting, synaptogenesis, and angiogenesis in the injured brain. These restorative processes are associated with improvement in neurological functional outcome. Administration of neurorestorative agents involve treatment of stroke at subacute (>24 h) time points, resulting in greater availability for treatment for all patients rather than the few, that is, approximately 5% of patients who are treated with the time-limited thrombolytic agent rt-PA (<4.5 h).[5] Preclinical data demonstrating neurorestoration are growing robustly, with many agents poised to be tested in the clinical trials.

Tβ4 is a pleiotropic peptide exhibiting many actions in several different types of tissues.[6] Tβ4 has anti-inflammatory activity and promotes angiogenesis in dermal wound healing and cardiac ischemia models. The fundamental property of Tβ4 is sequestration of G-actin monomers, which promotes cell migration by inhibiting actin-cytoskeletal organization.[7] Because of its G-actin–binding properties, Tβ4 promotes cardiomyocyte migration in models of cardiac infarction and keratinoctye

doi: 10.1111/j.1749-6632.2012.06651.x

migration in wound-healing models.[8,9] These properties support our hypothesis that Tβ4 is a neurorestorative agent. For example, after focal cerebral ischemia, the injured brain attempts to repair and remodel itself.[10] An important, well-documented regenerative response is the proliferation of neural progenitor cells (NPC) in the subventricular zone (SVZ) of the lateral ventricle after stroke.[11,12] The SVZ of the rat contains neural stem cells that produce neuroblasts that regularly migrate to the olfactory bulb via the rostral migratory stream and differentiate into granule neurons.[12,13] The replacement of olfactory neurons is critical to the survival of the rat. However, after experimental focal cerebral ischemia, the NPCs migrate out of the SVZ to the ischemic boundary regions to promote a process of repair.[13] Since actin dynamics play a critical role in cell migration and synaptogenesis, we propose that administration of Tβ4 may enhance this migration of progenitor cells to accelerate and potentiate recovery after stroke or other models of neurological injury.

Tβ4 is an agent that has been tested in three different models of neurological injury in our laboratory, and its small peptide properties (43 amino acids) make it an ideal candidate for neurorestorative therapy.[14–16] The first model is a rat embolic stroke model, which involves carefully placing a fibrin-rich blood clot into the middle cerebral artery (MCA) to induce stroke-like symptoms.[17] This model has the advantages of reproducibility, and more importantly, it is a clinically relevant model of human stroke. The second model is an experimental autoimmune encephalomyelitis (EAE) mouse model of multiple sclerosis, which is generated by immunization with myelin proteolipid protein (PLP) to induce neurological symptoms of paralysis, ataxia, and poor muscle tone.[15] Finally, our laboratory has a rat model of traumatic brain injury (TBI) in which injury is delivered to the anesthetized rat by impacting the cortex with a pneumatic piston.[16] All three models have the advantages of producing neurological deficits that can be measured over a long period of time. In this review, we describe observations of neurological improvement in each model of neurological injury as well as data supporting our hypothesis that Tβ4 is a neurorestorative agent, specifically promoting differentiation of OPCs into myelin-secreting oligodendrocytes (OLs), a process known as oligodendrogeneisis.

Embolic stroke model

The MCA of male Wistar rats (320 to 380 g, $n = 18$) was occluded by placement of an embolus at the origin of the MCA, as previously described.[17] Twenty-four hours after stroke, Tβ4 was administered intraperitonally (IP) and then every three days (6 mg/kg, IP) for four additional doses ($n = 9$). A battery of behavioral tests, including the adhesive-removal test (ART) and the modified Neurological Severity Score (mNSS), were performed before MCA occlusion and at 1, 7, 14, 21, 28, 35, 42, 49, and 56 days after MCAo by an investigator who

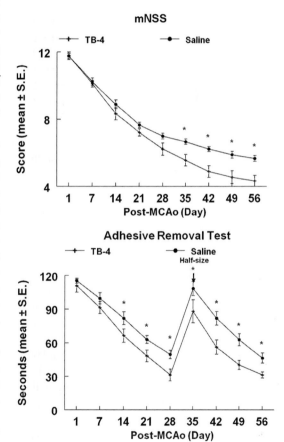

Figure 1. Embolic stroke rat model treated with Tβ4. The mNSS of embolic stroke rats ($n = 18$) treated with Tβ4 demonstrated a significant overall (treatment effect) improvement of neurological function ($P < 0.01$). The adhesive removal test of embolic stroke rats treated with Tβ4 also demonstrated a significant overall (treatment effect) improvement ($P < 0.01$). Significant effect ($P < 0.05$) at individual time points are indicated by asterisks. Adhesive-backed paper dots were reduced in size by one-half at day 35 (arrow) to increase sensitivity. Reprinted from Ref. 14, with permission from Elsevier.

was blinded to the experimental groups.[18] The ART measures the time it takes for the animal to remove sticky tabs from its paws and the mNSS measures motor, sensory, proprioception, and balance. Ischemic rats treated with saline ($n = 9$) were used as a control group. Animals were sacrificed after 56 days. Results (Fig. 1) from this experiment demonstrated that Tβ4-treated rats showed a 24.2% and a 29.9% overall improvement in the ART and mNSS scores (at time of sacrifice), respectively, when compared to controls (overall treatment effect, $P < 0.01$). Functional improvements persisted for at least 56 days after MCA occlusion. There were no significant differences of ischemic lesion volumes between the rats treated with Tβ4 (35.2% ± 6.7%) and with saline (33.1% ± 7.8%, $P > 0.05$), indicating that a neuroprotective mechanism was not responsible for the improvement. We, therefore, tested whether Tβ4 promotes axonal remodeling after stroke. Brain sections were stained using the Bielshowsky and Luxol fast blue staining to detect myelinated axons. Figure 2A demonstrates significant increases in staining area of myelinated axons in the striatal (white matter) ischemic boundary in the Tβ4 treatment group (215.3 ± 29.9%) when compared to the control group (115.2 ± 9.0%; $P < 0.05$). The increase of remyelination, which was associated with functional improvement, would suggest that cells that produce myelin, OLs and its precursors, and OPCs would be increased. We measured markers of OPCs, NG-2 (chondroitin sulfate proteoglycan), and OLs, CNPase (2′, 3′-cyclic nucleotide 3′-phosphodiesterase). Figure 2(B) and (C) demonstrate the expression of these two markers. When compared to controls, Tβ4 treatment significantly increased the density (cells/mm^2) of NG-2 positive cells in the SVZ (396.6 ± 19.6 vs. 209.1 ± 42.7) and striatum (130 ± 15.3 vs. 61.0 ± 7.6) ($P < 0.05$). NG-2 immunoreactivity was also increased in the corpus collosum (166.8 ± 26.0 vs. 78.3 ± 12.2, $P < 0.05$). CNPase area of increased staining was increased in the striatum (149.1% ± 9.4% vs. 115.2% ± 7.1%, $P < 0.05$). The association of improvement of neurological outcome and oligodendrogenesis supports our hypothesis of neurorestoration by Tβ4.

EAE model of multiple sclerosis

Our laboratory uses a standard mouse model of EAE.[15] EAE mice were administered saline ($n = 11$) or Tβ4 ($n = 10$) at a concentration of 6 mg/kg

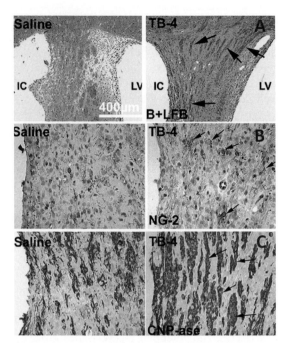

Figure 2. Embolic stroke rat model treated with Tβ4. The staining by Bielshowsky and Luxol fast blue (A) shows the myelin and axons in the white matter bundles of the striatum of saline- and Tβ4-treated rats (see arrows). There is a significantly increased density of Bielshowsky and Luxol fast blue staining in the Tβ4-treated rats compared to the demyelination of the saline control. LV, lateral ventricle; IC, ischemic core. NG-2 staining (B) is significantly increased in the ipsilateral SVZ and striatum adjacent to the ischemic core of Tβ4-treated rats when compared to saline control (see arrows). CNPase (C) is significantly increased in the striatum of Tβ4-treated rats when compared to saline control (see arrows). $P < 0.05$ for A, B, and C. Reprinted from Ref. 14, with permission from Elsevier.

IP on the day of PLP immunization, and then every three days (6 mg/kg) for four additional doses. Neurological function was scored using a standard scoring scale of 0–5. Mice were scored daily for clinical symptoms of EAE, as follows: 0, healthy; 1, loss of tail tone; 2, ataxia and/or paresis of hindlimbs; 3, paralysis of hind limbs and/or paresis of forelimbs; 4, tetraparalysis; 5, moribund or dead.[19] The higher score the more severe the disease. Results from this study showed that Tβ4 treatment improved neurological outcome nearly 50% when compared to controls (Fig. 3). Improvement was observed beginning at day 11 and extended to time of sacrifice at day 30. Similar to the embolic stroke model an increase in NG2 OPCs (447.7 ± 41.9 vs. 195.2 ± 31 cells/mm^2 in SVZ, 75.1 ± 4.7 vs.

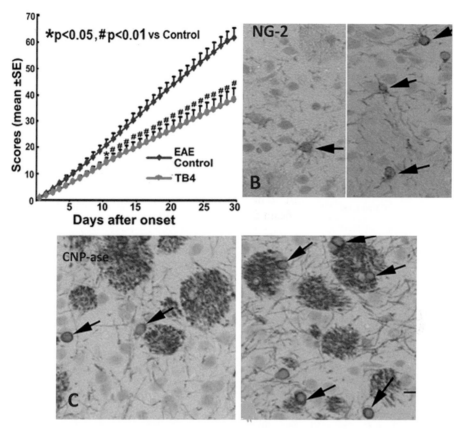

Figure 3. EAE mouse model of multiple sclerosis treated with Tβ4. The neurological response of EAE mice treated with or without Tβ4. The significant therapeutic Tβ4 effects were detected as early as day 11 after EAE onset. Nearly 50% relative functional recovery was observed in the Tβ4-treated group, compared to the saline controls with $P < 0.01$ for either the median score or the cumulative score up to 30 days. NG-2 cells (B) and CNPase cells (C) were significantly increased at 30 days after EAE onset in the Tβ4 treatment group compared to that in the saline group ($P < 0.05$). Reprinted from Ref. 15, with permission from Elsevier.

41.7 ± 3.2 cells/mm^2 in white matter) and CNPase$^+$ mature OLs staining area (267.5 ± 10.3 vs. 141.4 ± 22.9/mm^2) was also observed in the SVZ and white matter of the brain suggesting that oligodendrogenesis is occurring (Fig. 3B and C). Similar to the embolic stroke model, these results suggest an association of improvement of neurological outcome and oligodendrogenesis by Tβ4.

Traumatic brain injury model

A controlled cortical impact model was used by our laboratory to model TBI in rats.[20] Young adult male Wistar rats (330 gm) were anesthetized with chloral hydrate and placed in a stereotactic frame. Two 10-mm-diameter craniotomies were performed adjacent to the central suture. The contralateral craniotomy allowed for movement of cortical tissue laterally. The dura mater was kept intact over the

cortex. Injury was delivered by impacting the left (ipsilateral) cortex with a pneumatic piston containing a 6-mm-diameter tip at a rate of 4 m/sec and 2.5 mm of compression. Velocity was measured with a linear velocity displacement transducer. The animals were divided into three groups: (1) sham (surgery without TBI) group ($n = 6$); (2) surgery + TBI + saline group ($n = 9$); and 3) surgery + TBI + Tβ4 group ($n = 10$). Tβ4 was administered at a dose of 6 mg/kg IP beginning at day 1 after injury, and every three days for four additional doses until sacrifice at day 35. Neurological outcome was measured using three standardized tests: the Morris water maze test, the foot fault test, and the mNSS test, as previously described.[16] The Morris water maze test measures the animal's spatial learning impairments while swimming in a shallow pool. The foot fault test measures sensorimotor function by

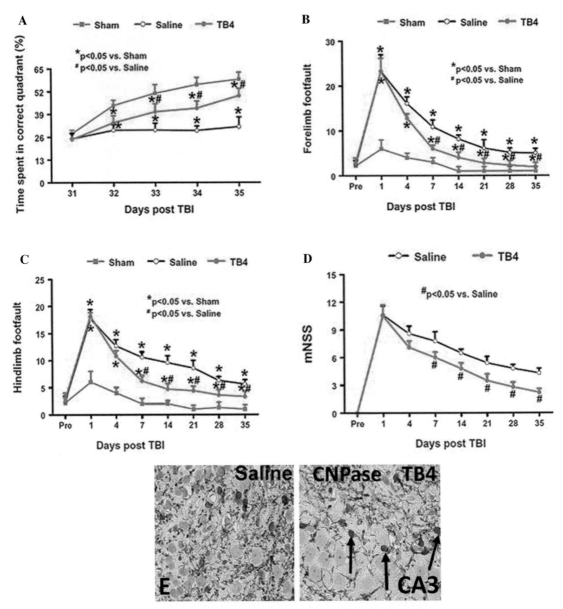

Figure 4. Rat model of traumatic brain injury treated with Tβ4. Tβ4 treatment improves spatial learning performance measured by the Morris water maze test at days 33–35 compared with the saline group (A). Tβ4 treatment significantly reduces forelimb foot faults at days 7–35 compared with the saline-treated group (B). Tβ4 treatment significantly reduces hindlimb foot faults at days 7–35 compared with the saline-treated group (C). Line graph showing the functional improvement detected on the mNSS (D). Treatment with Tβ4 significantly lowers mNSS at days 7–35 compared with the saline group. Pre = preinjury level. Treatment with Tβ4 significantly increases CNPase cells in the CA3 region of the hippocampus (E) ($P < 0.05$). Reprinted from Ref. 16, with permission from the JNS Publishing Group.

allowing the rat to walk on a wired mesh, with each paw that slips between the wires counting as a misstep. The results are shown in Figure 4A–D, Tβ4-treated rats showed reduced deficits in all three tests with a near 75% improvement in the Morris water maze test, and a 50% improvement in both the foot fault and mNSS.

Tβ4-mediated oligodendrogenesis was observed in the CA3 of the hippocampus (Fig. 4E). The hippocampus region participates in the organization of both short- and long-term memories and spatial orientation. This specific regional finding is associated with and may account for the reduced impairment of spatial learning test (Morris water maze). TBI by itself stimulates increased OPCs in all regions measured; however, TB4 only increased the more mature OL, as evidence by increased CNPase staining (120.5 ± 14 vs. 256 ± 24 cells/mm^2) in the CA3 region of the hippocampus ($P < 0.05$). This measurement was performed at day 35, and future studies are needed to determine if Tβ4 promotes oligodendrogenesis at earlier time points in either the cortex or denate gyrus.

Clinical translation

The central nervous system has the ability to regenerate damaged axonal connections.[4,21] The outgrowth of collateral sprouting resulting from plasticity of neurons creates new axonal connections and new circuitry in the injured brain.[10] Functional improvement occurs after stroke when surviving neurons undergo axonal sprouting and synaptogenesis.[22] After stroke, neuroblasts in the SVZ migrate toward the ischemic boundary regions, suggesting that the neurogenic response after stroke has the potential to be manipulated to increase the number of new migrating NPCs that possess the ability to differentiate into neurons and oligodendrocytes.[13] Increases in the proliferating progenitor cells in the ischemic brain may provide an opportunity to repair axonal connections.

The resulting improvement in neurological outcome after treatment with Tβ4 in our three models of neurological injury reflects a Tβ4-mediated neurorestorative process. A common observation in all three models is oligodendrogenesis and/or the production of mature myelin-secreting OL from OPC. Remyelination has been well studied in various adult animal models and involves the generation of new mature OLs,[23,24] which are derived from adult OPCs whose origins are from white matters and the SVZ. These mature OLs spread throughout the adult white matter. The general scientific consensus is that remyelination occurs only from OPCs and not from surviving OLs or from mature surviving OLs adjacent to the injured axons.[23–25] Mature OLs are, for the most part, unable to migrate or divide. OLs are highly vulnerable to focal cerebral ischemia[26] and other toxic factors resulting in neurological deficits. Our research suggests that Tβ4 is a

potential candidate as a neurorestorative agent. Tβ4 was tested in a randomized, double-blind, placebo-controlled, dose-response phase 1A and 1B study of the safety and tolerability of the intravenous administration of Tβ4 and its pharmacokinetics after single doses in healthy volunteers (RegeneRx Biopharmaceuticals, Inc., Rockville, MD).[27] The drug was found to be safe and well tolerated. Therefore, our preclinical results along with the clinical safety trial suggest that clinical trials using Tβ4 should be considered.

Conclusion

The three models of neurological injury treated with Tβ4 described in this review support our hypothesis that Tβ4 is a potential neurorestorative agent. Neurorestorative agents target intact parenchymal cells to promote brain remodeling or repair of damaged tissues. Improvement of neurological functional outcome observed in all three models is associated with oligodendrogenesis and/or the differentiation of OPC into mature OL. Tβ4 could, in theory, treat all three diseases, stroke, multiple sclerosis, and TBI. Presently, research is focusing on optimizing the dose and the time to administer the peptide after injury. These preclinical studies suggest that clinical trials are warranted for Tβ4 treatment of these debilitating diseases in humans.

Acknowledgments

This research was supported in part by NINDS and NIA Grants PO1 NS23393, RO1 NS062832, R01 NS075156, and R01 AG038648. Tβ4 was supplied by RegeneRx Biopharmaceuticals Inc. under a Material Transfer Agreement. A U.S. Provisional Patent 61/163,556 has been filed for use of Tβ4 in neurological disease.

Conflicts of interest

The authors declare no conflicts of interest.

References

1. Sharma, J.C., S. Fletcher & M. Vassallo. 1999. Strokes in the elderly – higher acute and 3-month mortality – an explanation. *Cerebrovasc. Dis.* **9:** 2–9.
2. Reitz, C., C. Brayne & R. Mayeux. 2011. Epidemiology of Alzheimer disease. *Nat. Rev. Neurol.* **7:** 137–152.
3. Fuchs, V.R. 2008. Three "inconvenient truths" about health care. *N. Engl. J. Med.* **359:** 1749–1751.
4. Zhang, Z.G. & M. Chopp. 2009. Neurorestorative therapies for stroke: underlying mechanisms and translation to the clinic. *Lancet Neurol.* **8:** 491–500.

5. Hacke, W. *et al.* 2008. Thrombolysis with alteplase 3 to 4.5 hours after acute ischemic stroke. *N. Engl. J. Med.* **359:** 1317–1329.

6. Goldstein, A.L., E. Hannappel & H.K. Kleinman. 2005. Thymosin β4: actin-sequestering protein moonlights to repair injured tissues. *Trends Mol. Med.* **11:** 421–429.

7. Huff, T. *et al.* 2001. β-Thymosins, small acidic peptides with multiple functions. *Int. J. Biochem. Cell Biol.* **33:** 205–220.

8. Smart, N. *et al.* 2007. Thymosin β4 induces adult epicardial progenitor mobilization and neovascularization. *Nature* **445:** 177–182.

9. Malinda, K.M. *et al.* 1999. Thymosin β4 accelerates wound healing. *J. Invest. Dermatol.* **113:** 364–368.

10. Cramer, S.C. & M. Chopp. 2000. Recovery recapitulates ontogeny. *Trends Neurosci.* **23:** 265–271.

11. Alvarez-Buylla, A. & J.M. Garcia-Verdugo. 2002. Neurogenesis in adult subventricular zone. *J. Neurosci.* **22:** 629–634.

12. Arvidsson, A. *et al.* 2002. Neuronal replacement from endogenous precursors in the adult brain after stroke. *Nat. Med.* **8:** 963–970.

13. Zhang, R. *et al.* 2004. Activated neural stem cells contribute to stroke-induced neurogenesis and neuroblast migration toward the infarct boundary in adult rats. *J. Cereb. Blood Flow Metab.* **24:** 441–448.

14. Morris, D.C. *et al.* 2010. Thymosin β4 improves functional neurological outcome in a rat model of embolic stroke. *Neuroscience* **169:** 674–682.

15. Zhang, J. *et al.* 2009. Neurological functional recovery after thymosin β4 treatment in mice with experimental auto encephalomyelitis. *Neuroscience* **164:** 1887–1893.

16. Xiong, Y. *et al.* 2011. Treatment of traumatic brain injury with thymosin β(4) in rats. *J. Neurosurg.* **114:** 102–115.

17. Zhang, R. *et al.* 1997. A rat model of focal embolic cerebral ischemia. *Brain Res.* **776:** 83–92.

18. Cenci, M.A., I.Q. Whishaw & T. Schallert. 2002. Animal models of neurological deficits: how relevant is the rat? *Nat. Rev. Neurosci.* **3:** 574–579.

19. Zhang, J. *et al.* 2005. Erythropoietin treatment improves neurological functional recovery in EAE mice. *Brain Res.* **1034:** 34–39.

20. Mahmood, A., D. Lu & M. Chopp. 2004. Marrow stromal cell transplantation after traumatic brain injury promotes cellular proliferation within the brain. *Neurosurgery* **55:** 1185–1193.

21. Hilliard, M.A. 2009. Axonal degeneration and regeneration: a mechanistic tug-of-war. *J. Neurochem.* **108:** 23–32.

22. Chopp, M., Y. Li & J. Zhang. 2008. Plasticity and remodeling of brain. *J. Neurol. Sci.* **265:** 97–101.

23. Franklin, R.J. & C. Ffrench-Constant. 2008. Remyelination in the CNS: from biology to therapy. *Nat. Rev. Neurosci.* **9:** 839–855.

24. Nait-Oumesmar, B. *et al.* 2008. The role of SVZ-derived neural precursors in demyelinating diseases: from animal models to multiple sclerosis. *J. Neurol. Sci.* **265:** 26–31.

25. Franklin, R.J. 2002. Why does remyelination fail in multiple sclerosis? *Nat. Rev. Neurosci.* **3:** 705–714.

26. Pantoni, L., J.H. Garcia & J.A. Gutierrez. 1996. Cerebral white matter is highly vulnerable to ischemia. *Stroke* **27:** 1641–1646; discussion 1647.

27. Ruff, D. *et al.* 2010. A randomized, placebo-controlled, single and multiple dose study of intravenous thymosin β4 in healthy volunteers. *Ann. N.Y. Acad. Sci.* **1194:** 223–229.

Ann. N.Y. Acad. Sci. ISSN 0077-8923

ANNALS OF THE NEW YORK ACADEMY OF SCIENCES

Issue: *Thymosins in Health and Disease*

Therapeutic potential of thymosin β4 in myocardial infarct and heart failure

Christoffer Stark,[1,2] Pekka Taimen,[3] Miikka Tarkia,[4] Jussi Pärkkä,[4] Antti Saraste,[4] Tero-Pekka Alastalo,[5] Timo Savunen,[1] and Juha Koskenvuo[2]

[1]Department of Surgery, Turku University Central Hospital, Turku, Finland. [2]Research Center of Applied and Preventive Cardiovascular Medicine, University of Turku, Turku, Finland. [3]Department of Pathology, University of Turku and Turku University Hospital, Turku, Finland. [4]Turku PET Center, University of Turku, Turku, Finland. [5]Department of Pediatrics, Helsinki University Hospital, Helsinki, Finland

Address for correspondence: Dr. Christoffer Stark, Research Centre of Applied and Preventive Cardiovascular Medicine, University of Turku, Kiinamyllynkatu 10, 20520 Turku, Finland. christoffer.stark@utu.fi

Thymosin β4 (Tβ4) is a peptide known for its abilities to protect and facilitate regeneration in a number of tissues following injury. Its cardioprotective effects have been evaluated in different animal models and, currently, a clinical trial is being planned in patients suffering from acute myocardial infarction. This paper focuses on the effects of Tβ4 on cardiac function in animal studies utilizing different imaging modalities for outcome measurements.

Keywords: thymosin beta 4; heart; myocardial infarction; cardioprotection; imaging

Myocardial infarction and heart failure

Ischemic heart disease is a growing concern worldwide. The prevalence of conditions such as heart failure is expected to grow by more than fifty percent until the year 2030.[1] Acute myocardial infarction (AMI) usually results from atherosclerotic plaque rupture and thrombosis of the coronary artery.[2] Immediate treatment for this condition relies on protective strategies to lower myocardial oxygen consumption and antithrombotic therapy together with early revascularization.[3] Treatments for myocardial infarction have improved greatly during the last decades with more patients surviving the acute phase.[4] There is, however, no current therapy replacing already lost, scarred myocardium. The scarred myocardium leads to morphological and functional changes in the heart that clinically manifests as arrhythmias and congestive heart failure (CHF). The pathophysiology of this condition is not yet fully understood, but it is partly explained by hypertrophy and fibrosis of the heart, along with activation of the neurohumoral system.[5] CHF is a chronic illness associated with high hospitalization rates and decreased quality of life. The five-year survival rate is less than 50% in advanced CHF.[6]

Thymosin β4 in heart disease

Thymosin β4 (Tβ4) is a naturally occurring G-actin–sequestering peptide that has abundant protective effects on different cell types after various types of cell injuries both *in vitro* and *in vivo*.[7] The peptide is expressed in all studied cell types, except red blood cells. High concentrations have been measured in white blood cells and in platelets, from which, the peptide is secreted during tissue damage and thrombosis formation.[8] At the sites of thrombosis, Tβ4 is cross-linked to fibrin and collagen through activated factor XIII.[9] Kaur *et al.* showed that low concentrations of Tβ4 (0.2–0.5 mcM) increase activated platelet adhesion rate and number in a flow-chamber model. With higher doses of Tβ4 (>1 mcM), the adhesion rate returned to normal. They also found interactions with fibrinogen and speculated that higher concentrations of Tβ4 could attenuate platelet deposition.[10] The influence of Tβ4 on blood coagulation is poorly understood and needs to be further clarified, especially in thromboembolic conditions such as AMI.

Endoproteinase-mediated cleavage of Tβ4 leads to the release of the antifibrotic and proangiogenic tetrapeptide, Ac-SDKP, duplicating the angiogenic

doi: 10.1111/j.1749-6632.2012.06695.x

Table 1. Effects of Tβ4 on cardiac function

Animal model	Tβ4 treatment	Imaging modality	Outcome measure	Finding	Reference
Mouse MI	400 ng × 1 i.c. or 150 µg i.p. every third day or both i.c. and i.p.	Echo two and four weeks post-MI	EF, FS, LVEDD, LVESD	Increased systolic function (EF 57.7% vs. 28.2% in controls) Decrease in LV systolic and diastolic diameters	14
Mouse MI	200 µg i.p. one, two, three, five, and seven days post-MI	Echo seven days post-MI	EF, FS, LV mass, LVEDD	Increase in systolic function (EF 52% vs. 14% in controls) Less LV hypertrophy and lower diastolic diameter	32
Mouse MI	150 µg i.p. daily for two weeks or 400 ng i.c., three weekly doses post-MI	Echo two days and four weeks post-MI	EF, LVESV, LVEDV, infarct size	Slight improvement in EF Reduced LV dilatation	Stark *et al.*, unpublished data
Mouse MI	12 mg/kg i.p. every other day post-MI	CMR 7, 14, and 28 days post-MI	EF, LVEDV, LVESV, infarct size	Increase in EF (41% vs. 23% in controls), less LV dilatation	23
Rat MI	PEG hydrogels 40 µg/mL, 60 µL i.c. post-MI	CMR three days and six weeks post-MI	EF, LVESV, LVEDV, infarct size	Minimal improvement compared to hydrogels alone	33
Pig I-R	15 mg retroinfused into left interventricular vein postischemia	Sonomicrometry 24 h post-I-R	Subendocardial segment shortening	Increase in SES during rest and pacing	15
			Dp/dtmax	Dp/dtmax increased at rapid pacing	
Mouse Mdx	Dystrophin deficiency; 150 µg every other day for six months	Echo at 6 months	FS, LVESV, LVEDV, LV mass, TDE	No differences between groups	36

Continued

Table 1. *Continued*

Animal model	Tβ4 treatment	Imaging modality	Outcome measure	Finding	Reference
Mouse	TB4 knockout versus wild-type	Echo at 12, 24, 36, and 52 weeks	EF, FS, LVEDD, LVESD	No differences between groups	18

MI, myocardial infarction; i.p., intraperitoneal; i.c, intramyocardial; EF, left ventricular ejection fraction; FS, fractional shortening; LVEDD, left ventricle end-diastolic diameter; LVESD, left ventricle end-systolic diameter; LV, left ventricle; LVESV, left ventricle end-systolic volume; LVEDV, left ventricle end-diastolic volume; CMR, cardiac magnetic resonance; Dp/dtmax, systolic force.

properties shared between Ac-SDKP and its known sole precursor, Tβ4.[11] Ac-SDKP is further cleaved by angiotensin converting enzyme (ACE). ACE-inhibitors have been reported to be cardioprotective. In the presence of ACE-inhibitors the concentration of Ac-SDKP increases in blood, and it is possible that at least some of the cardioprotective effects of ACE-inhibitors are mediated by Ac-SDKP.[12]

In the animal models for AMI, Tβ4 is known to increase cardiomyocyte survival, angiogenesis, and cell migration. In addition, it reduces inflammation, fibrosis, and apoptosis.[13–17] Tβ4 is present during embryonic heart development. By knocking down Tβ4 using shRNA, Smart *et al.* reported an essential role for Tβ4 in regulating cardiac vessel development.[13] In a more recent study conducted on Tβ4-knockout mice, the animals expressed normal vascular and cardiac phenotypes, indicating a smaller developmental role for Tβ4.[18] Bock-Marquette *et al.* discovered that Tβ4 interacts with the ILK-PINCH complex, leading to the activation of survival kinase Akt, which is an important regulator of several fundamental cellular mechanisms related to injury response and tissue healing.[14] Tβ4 also induces PKC mediated angiogenesis in the heart, possibly through VEGF signaling.[16] Furthermore, Tβ4 promotes migration of endothelial cells, partly by acting through purinergic receptors.[19,20] In models of dermal, corneal eye, and cardiac injury, and endotoxin-induced sepsis, Tβ4 has been shown to downregulate inflammatory mediators and reduce inflammatory cell infiltrates.[11] Part of the anti-inflammatory effect of Tβ4 is mediated by downregulation of the cytokine NF-κB, although cardioprotective findings have not been confirmed *in vivo*.[21] Tβ4 reduces apoptosis and induces survival genes. For example, Tβ4 has been shown to inhibit apoptosis in endothelial cells and activate cell survival signaling pathways in cardiomyocytes. Following myocardial infarction the number of apoptotic cells peaks at 24 h, which is when the antiapoptotic effects of Tβ4 have been observed. In cardiac fibroblasts, Tβ4 decreases the expression of several profibrotic genes after experimental oxidative stress by upregulating antioxidant genes.[22] Given these positive effects, Tβ4 is thought to decrease the amount of lost myocardium following MI and thereby prevent the progression of heart failure.

Previously, Smart *et al.* reported that epicardial-derived cardiac progenitor cells were integrated both structurally and functionally with resident cardiomyocytes after MI, suggesting a source of new heart muscle.[23] In this work, Tβ4 was administered before the onset of MI, which is unpractical in clinical situations. In a similar study by Zhou *et al.*, the authors failed to show a cardiomyocyte fate of epicardial cells when Tβ4 was administered post-MI. They did, however, report activation of epicardial cells and cardioprotective effects, supporting earlier findings.[24] In adults, cells derived from the epicardium most likely represent smooth muscle cells and fibroblasts, and it is currently uncertain whether *de novo* cardiomoycytes are derived from the epicardium.[25] Other possible sources of newly formed cardiomyocytes are circulating stem cells and cardiac interstitial cells that undergo transdifferentiation after appropriate stimuli.[26] The significance and clinical relevance of regenerating cardiomyocytes are, however, unclear due to small number of activated cells.

Following AMI and resuscitation in a pig model, levels of endogenous Tβ4 were shown to increase significantly.[27] Under native conditions, Tβ4 and

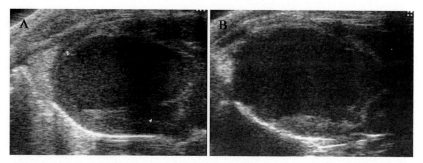

Figure 1. Parasternal long-axis B-mode echocardiography of left ventricle at two days (A) and four weeks (B) post-MI in mouse. Ventricular remodeling is seen as dilatation, wall thinning, and infarct expansion. Data represent work in progress.

other paracrine factors are upregulated in the epicardial layer of heart following MI, peaking at day five. The epicardial changes seen are thought to be involved in endogenous protective and reparative processes and augmenting these effects could be an attractive target in regenerating and protecting injured myocardium.[28] A study conducted on patients with severe coronary artery disease showed a significant correlation between good collateral vessel development and higher serum levels of Tβ4.[29] So far, data from clinical trials on cardioprotection of Tβ4 are lacking. In healthy subjects, Tβ4 administration is shown to be safe and well tolerated at doses up to 1,260 mg/day.[30] Therefore, it is considered that Tβ4 holds great promise in improving the care of patients suffering from ischemic heart disease.

Functional studies with animal models

Characterization of novel biological treatments requires initial research using animal models. In experimental cardiac research, large animal models and especially pigs are considered superior to other animals due to similarities to human cardiac anatomy and physiology. These similarities improve translation of research data into clinical practice. Pig models for chronic myocardial ischemia are, however, difficult and expensive to produce and maintain, which makes research with rodent models more appealing.[31]

As mentioned before, Tβ4 has been shown to be important during embryonic morphogenesis of the heart. Banerjee *et al.* challenged this view and investigated whether global knockout of Tβ4 influenced cardiac function and morphology as measured by echocardiography in 12- to 52-week-old mice. The authors observed no differences between knockout and wild-type mice and concluded that Tβ4 is dis-

pensable for heart development and adult cardiac function.[18]

There are numerous studies indicating beneficial effects of Tβ4 on cardiac cells after experimental damage.[7] Few studies, though, have shown improvement in cardiac function after ischemic insult (Table 1). The great majority of the animal models rely on chronic MI in mouse or rat, in which the left anterior descending coronary artery (LAD) is occluded permanently by external ligation.[14,23,32,33] This model offers reproducible infarct sizes in skillful hands due to the small variation in LAD anatomy.[34] It is, however, important to bear in mind that there are significant strain and species differences in normal heart function and also in reactions to ischemia and healing processes of myocardial infarction.[35]

The administration of Tβ4 via intraperitoneal or intramyocardial injection after myocardial infarction has shown a remarkable improvement in left ventricular function and a decrease in left ventricular remodeling.[14,23,32,33] Similar results have been obtained with both routes of administration, although our results have shown better cardioprotective effects after direct intramyocardial treatment (unpublished results). Therapeutic doses of Tβ4 used range from 400 ng as single injections to several injections of 150 μg or more depending on respective study protocols. There are currently no clear dosing strategies favoring one over the other when preventing LV remodeling.

To date, the functional effects of Tβ4 have been studied using echocardiography and cardiac magnetic resonance imaging (CMR).[14,15,18,23,32,33,36] Echocardiography is user dependent and, in the case of small animal models, also equipment dependent. Attaining high-resolution and high-quality images

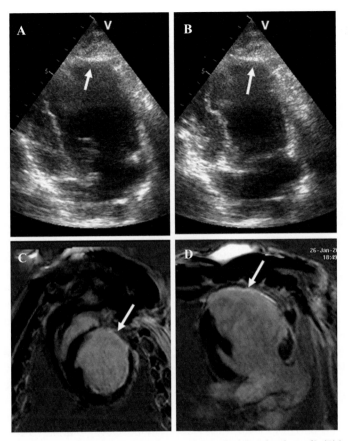

Figure 2. Echocardiography four weeks after MI in a pig model. End-diastolic (A) and end-systolic (B) images show akinesia and hyperdensity in the anteroseptal region (arrows). After two months, there is clear thinning and fibrosis of the infarcted area as seen in CMR short axis (C) and long axis (D) views (arrows). Data represent work in progress.

requires sophisticated hardware and software designed for preclinical research. Factors such as temperature and depth of anesthesia can influence cardiac function and lead to misinterpretations especially in rodents.[37] CMR is more time consuming and expensive, and requires synchronization with respiration and heart rhythm. The spatial resolution is, however, better compared to images acquired with echocardiography and therefore, function and anatomy can be measured more accurately. Assessment of myocardial viability can be made with pharmacological stress echocardiography or CMR, the latter also offering the use of contrast enhancement in distinguishing viable from nonviable myocardium.[38]

The functional parameters reported in animal studies on Tβ4 vary in type and extent. Most commonly used are fractional shortening (FS) and left ventricular ejection fraction (LVEF), which is the amount of blood ejected from the left ventricle during one systolic contraction. Although load dependent, LVEF is a powerful tool in assessing global dynamics of the left ventricle and predicts worse outcomes in patients suffering from advanced heart failure.[39] After myocardial infarction in Tβ4-treated rodents, EF is reported to be increased by 18–30% at four weeks when compared to controls.[14,23,32] In our pilot study, we observed a slight increase in EF in intramyocardially treated animals when compared at two days and four weeks post-MI. In controls, EF was decreased by 10%, indicating a progressive LV remodeling. Left ventricular internal diameters during systole and diastole (LVESD and LVEDD) can be misleading in the infarcted heart, due to regional differences in contraction especially when utilizing infarct models. Therefore, ventricular volumes (LVESV and LVEDV) calculated from biplane 2D (Fig. 1) or 3D images should be used. In chronic

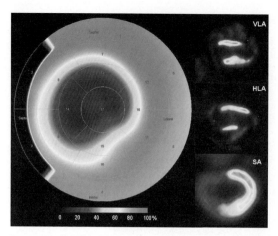

Figure 3. [¹⁸F]FDG PET and myocardial viability two months post-MI in pig. Tracer uptake is missing in the infarcted anteroseptal area and reduced in the peri-infarct zone. VLA = vertical long axis view, HLA = horizontal long axis view, SA = short axis view. Data represent work in progress.

heart failure, end diastolic and systolic volumes of the left ventricle are good predictors of outcome and reverse remodeling.[40] Other parameters, such as LV mass and wall thickness, can also be used to determine remodeling. Tβ4 has been shown to dramatically prevent early changes in left ventricle geometry after MI.[14,23,32]

Follow-up times in most studies using small animal models have ranged from one to six weeks.

A six-month follow-up period was used in a study conducted on dystrophin deficient *mdx* mice, the mouse model of Duchenne muscular dystrophy with chronic/progressive heart failure. In the latter study, Tβ4 was administered every second day intraperitoneally and echocardiography was applied to measure functional and geometric outcome.[36] No difference between the Tβ4-treated and control animals was observed in this study. In ischemic models, the protective effects have been visible already at one and two weeks postinfarction, indicating early protective effects of Tβ4. CHF is, however, a progressive condition with ongoing remodeling of the heart and beneficial effects at early time points do not necessarily reflect long-term outcomes.

In a large-animal model, Hinkel *et al.* subjected pigs to experimental ischemia-reperfusion injury by balloon occlusion of the LAD for 60 minutes.[15] Tβ4 was given through retroinfusion into the accompanying vein at the end of ischemia and cardiac function was assessed at 24-h postreperfusion. Subendocardial segmental shortening assessed by sonomicrometry showed improved functional reserve of the ischemic myocardium during rest and rapid pacing in treated animals. This is, to date, the only study conducted on a large-animal model. Our group is currently working with a pig model for chronic ischemia and heart failure, where the LAD is gradually occluded by an ameroid

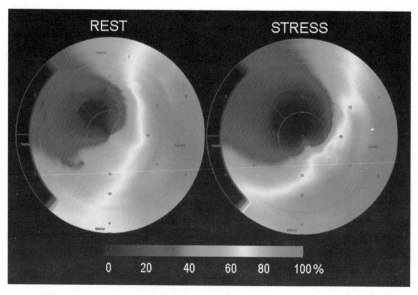

Figure 4. Radiowater PET myocardial perfusion during rest and adenosine stress in a pig model of MI. Data represent work in progress.

constrictor. Total occlusion of the artery occurs within three weeks after placement of the constrictor. This increases survival rates when compared to direct occlusion. Initial results show good disease progression and this will be an interesting model for further investigations on Tβ4-mediated cardioprotection (Fig. 2).

Other imaging modalities in assessing Tβ4-mediated cardioprotection

In previous studies, mechanical values have been used as end-points in describing the cardioprotective effects of Tβ4. With the use of new imaging modalities, it is possible to attain mechanistic insight into the positive effects mediated by Tβ4 *in vivo*.[41] Interstitial fibrosis is an important step in the development of CHF. The antifibrotic properties of Tβ4 can be quantified using CMR-based diffuse fibrosis imaging and possibly by positron emission tomography (PET), although this method is not yet validated in a large-animal model.[42,43] PET/SPECT imaging can also be used for evaluating metabolic changes, myocardial perfusion, and different markers for inflammation, angiogenesis, and apoptosis (Fig. 3). Furthermore, PET imaging allows quantitative measurement of perfusion (Fig. 4).[44]

Conclusions

There is strong preclinical evidence that Tβ4 treatment is beneficial after myocardial infarction, possibly by attenuating ischemia-reperfusion injury. Further research is however needed, especially in large animal models and human trials. The effects seem to be mainly cardioprotective, although there is some evidence for enhanced cardiac repair as well. The dosing and route of administration of Tβ4 are issues still warranting answers. Little is also known about the effect on long-term outcomes and so called hard end-points after myocardial injury, these being the most important measures in any biological treatment for critical conditions such as AMI or CHF. Results gathered by using conventional imaging techniques can be difficult to interpret and translate into clinical practice because there are no powerful prognostic values available for indicating positive outcome in treatments for mild or developing CHF. Quantitative imaging of fibrosis could be a good tool for assessment of the severity and prognosis of CHF. This could also be helpful in evaluating the thera-

peutic potential of new pharmacological agents such as Tβ4.

Conflicts of interest

The authors declare no conflicts of interest.

References

1. Heidenreich, P., J. Trogdon, O. Khavjou, *et al.* 2011. Forecasting the Future of Cardiovascular Disease in the United States: a Policy Statement From the American Heart Association. *Circulation* **123**: 933–944.
2. Fuster, V., L. Badimon, J.J. Badimon, *et al.* 1992. The pathogenesis of coronary artery disease and the acute coronary syndromes (1–2). *N. Engl. J. Med.* **326**: 242–250.
3. Van de Werf, F., J. Bax, A. Betriu, *et al.* 2008. Management of acute myocardial infarction in patients presenting with persistent ST-segment elevation: the Task Force on the Management of ST-Segment Elevation Acute Myocardial Infarction of the European Society of Cardiology. *Eur. Heart J.* **29**: 2909–2945.
4. Fox, K., P. Steg, K. Eagle, *et al.* 2007. Decline in rates of death and Heart failure in acute coronary syndromes, 1999–2006. *JAMA* **297**: 1892–1900.
5. Cohn, J.N., R. Ferrari & N. Sharpe. 2000. Cardiac remodeling—concepts and clinical implications: a consensus paper from an international forum on cardiac remodeling. *JACC* **35**: 569–582.
6. Ketchum, E.S. & W.C. Levy. 2011. Multivariate risk scores and patient outcomes in advanced Heart failure. *Congest Heart Fail* **17**: 205–212.
7. Goldstein, A., E. Hannappel, *et al.* 2012. Thymosin b4: a multi-functional regenerative peptide. Basic properties and clinical applications. *Expert Opin. Biol. Ther.* **12**: 37–51.
8. Hannappel, E. & M. van Kampen. 1987. Determination of Thymosin β 4 in human blood cells and serum. *J. Chromatogr.* **397**: 279–285.
9. Huff, T., A. Otto, *et al.* 2002. Thymosin β 4 is released from human blood platelets and attached by factor XIIIa (transglutaminase) to fibrin and collagen. *FASEB J.* **16**: 691–696.
10. Kaur, H., R. Heeney, *et al.* 2010. Whole blood, flow-chamber studies in real-time indicate a biphasic role for thymosin β-4 in platelet adhesion. *Biochemica et Biophysica Acta* **1800**: 1256–1261.
11. Crockford, D., N. Turjman, C. Allan & J. Angel. 2010. Thymosin β4: structure, function, and biological properties supporting current and future clinical applications. *Ann. N.Y. Acad. Sci.* **1194**: 179–189.
12. Hannappel, E. 2010. Thymosin β4 and its posttranslational modifications. *Ann. N.Y. Acad. Sci.* **1194**: 27–35.
13. Smart, N., C.A. Risebro, A.A. Melville, *et al.* 2007. Thymosin β4 induces adult epicardial progenitor mobilization and neovascularization. *Nature* **445**: 177–182.
14. Bock-Marquette, I., A. Saxena, M.D. White, *et al.* 2004. Thymosin β4 activates integrin-linked kinase and promotes cardiac cell migration, survival and cardiac repair. *Nature* **432**: 466–472.

15. Hinkel, R., C. El-Aouni, T. Olson, *et al.* 2008. Thymosin β4 is an essential paracrine factor of embryonic endothelial progenitor cell-mediated cardioprotection. *Circulation* **117:** 2232–2240.

16. Bock-Marquette, I., S. Srivasatava, G.C. Pipes, *et al.* 2009. Thymosin β4 mediated PKC activation is essential to initiate the embryonic coronary developmental program and epicardial progenitor cell activation in adult mice in vivo. *J. Mol. Cell Cardiol.* **46:** 728–738.

17. Shrivastava, S., D. Srivastava, E.N. Olson, *et al.* 2010. Thymosin β4 and cardiac repair. *Ann. N.Y. Acad. Sci.* **1194:** 87–96.

18. Banerjee, I. & J. Zhang. 2012. Thymosin β 4 Is dispensable for murine cardiac development and function. *Circ. Res.* **110:** 456–464.

19. Smart, N., A. Rossdeutsch & P. Riley. 2007. Thymosin b4 and angiogenesis: modes of action and therapeutic potential. *Angiogenesis* **10:** 229–241.

20. Freeman, K.W., B.R. Bowman & B.R. Zetter. 2011. Regenerative protein thymosin β-4 is a novel regulator of purinergic signaling. *FASEB J.* **25:** 907–915.

21. Rossdeutsch, A., N. Smart & P.R. Riley. 2008. Thymosin β4 and Ac-SDKP: tools to mend a broken heart. *J. Mol. Med.* **86:** 29–35.

22. Kumar, S. & S. Gupta. 2011. Thymosin β 4 prevents oxidative stress by targeting antioxidant and anti-apoptotic genes in cardiac fibroblasts. *PLoS One* **6:** e26912.

23. Smart, N., S. Bollini, K.N. Dube, *et al.* 2011. De novo cardiomyocytes from within the activated adult heart after injury. *Nature* **474:** 640–644.

24. Zhou, B., L.B. Honor, Q. Ma, *et al.* 2011. Thymosin β 4 treatment after myocardial infarction does not reprogram epicardial cells into cardiomyocytes. *J. Mol. Cell Cardiol.* **52:** 43–47.

25. Kispert, A. 2011. No muscle for a damaged heart: thymosin β 4 treatment after myocardial infarction does not induce myocardial differentiation of epicardial cells. *J. Mol. Cell Cardiol.* **52:** 10–12.

26. Bernstein, H. & D. Srivastava. 2012. Stem cell therapy for cardiac disease. *Pediatr. Res.* **71:** 491–499.

27. Shah, A. & S. Youngquist. 2011. Markers of progenitor cell recruitment and differentiation rise early during ischemia and continue during resuscitation in a porcine acute ischemia model. *J. Interferon Cytokine Res.* **31:** 6.

28. Zhou, B., L.B. Honor, H. He, *et al.* 2011. Adult mouse epicardium modulates myocardial injury by secreting paracrine factors. *J. Clin. Invest.* **121:** 1894–1904.

29. Bicer, A., Ö. Karakurt, *et al.* 2011. Thymosin β 4 is associated with collateral development in coronary artery disease. *Scandinavian J. Clin.Lab. Invest.* **71:** 625–630.

30. Ruff, D., D. Crockford, G. Girardi & Y. Zhang. 2010. A randomized, placebo-controlled, single and multiple dose study of intravenous thymosin β4 in healthy volunteers. *Ann. N.Y. Acad. Sci.* **1194:** 223–229.

31. Hausenloy, D., G. Baxter, *et al.* 2010. Translating novel strategies for cardioprotection: the Hatter Workshop Recommendations. *Basic Res. Cardiol.* **105:** 677–686.

32. Sopko, N., Y. Qin, A. Finan, *et al.* 2011. Significance of thymosin β4 and implication of PINCH-1-ILK-alpha-Parvin (PIP) complex in human dilated cardiomyopathy. *PLoS One* **6:** e20184.

33. Kraehenbuehl, T.P., L.S. Ferreira, A.M. Hayward, *et al.* 2011. Human embryonic stem cell-derived microvascular grafts for cardiac tissue preservation after myocardial infarction. *Biomaterials* **32:** 1102–1109.

34. Salto-Tellez, M., S. Lim, *et al.* 2003. Myocardial infarction in the C57BL/6J mouse, a quantifiable and highly reproducible experimental model. *Cardiovasc. Pathol.* **13:** 91–97.

35. Barnabei, M.S., N.J. Palpant & J.M. Metzger. 2010. Influence of genetic background on ex vivo and in vivo cardiac function in several commonly used inbred mouse strains. *Physiol. Genom.* **42A:** 103–113.

36. Spurney, C., H. Cha, *et al.* 2010. Evaluation of skeletal and cardiac muscle function after chronic administration of thymosin b-4 in the dystrophin deficient mouse. *PLoS One* **5:** e8976.

37. Ram, R. & D. Mickelsen. 2011. New approaches in small animal echocardiography: imaging the sounds of silence. *Am. J. Physiol. Heart Circ. Physiol.* **301:** H1765–H1780.

38. Tomlinson, D. & H. Becher. 2008. Assessment of myocardial viability: comparison of echocardiography versus cardiac magnetic resonance imaging in the current era. *Heart Lung Circ.* **17:** 173–185.

39. Solomon, S.D., N. Anavekar, H. Skali, *et al.* 2005. Influence of ejection fraction on cardiovascular outcomes in a broad spectrum of heart failure patients. *Circulation* **112:** 3738–3744.

40. Grayburn, P.A., C.P. Appleton, A.N. DeMaria, *et al.* 2005. Echocardiographic predictors of morbidity and mortality in patients with advanced heart failure: the β-blocker Evaluation of Survival Trial (BEST). *J. Am. Coll. Cardiol.* **45:** 1064–1071.

41. Shah, J., G. Fonarow, *et al.* 2011. Phase II trials in heart failure: the role of cardiovascular imaging. *Am. Heart J.* **162:** 3–15.e3.

42. Iles, L., H. Pfluger, A. Phrommintikul, *et al.* 2008. Evaluation of diffuse myocardial fibrosis in heart failure with cardiac magnetic resonance contrast-enhanced T1 mapping. *J. Am. Coll. Cardiol.* **52:** 1574–1580.

43. Aalto, K., A. Autio & E.A. Kiss. 2011. Siglec-9 is a novel leukocyte ligand for vascular adhesion protein-1 and can be used in PET imaging of inflammation and cancer. *Blood* **118:** 3725–3733.

44. Knuuti, J. & S. Kajander, *et al.* 2009. Quantification of myocardial blood flow will reform the detection of CAD. *J. Nucl. Cardiol.* **16:** 497–506.

Ann. N.Y. Acad. Sci. ISSN 0077-8923

ANNALS OF THE NEW YORK ACADEMY OF SCIENCES
Issue: *Thymosins in Health and Disease*

Thymosin β4 mobilizes mesothelial cells for blood vessel repair

Elaine L. Shelton and David M. Bader

The Stahlman Cardiovascular Research Laboratories, Program for Developmental Biology and Department of Medicine, Vanderbilt University Medical Center, Nashville, Tennessee

Address for correspondence: David M. Bader, Department of Medicine, Vanderbilt University, 2220 Pierce Avenue, 348 PRB, Nashville, TN 37232. david.bader@vanderbilt.edu

Mesothelium is the simple squamous epithelium covering all abdominal organs and the coeloms in which those organs reside. While the structural characteristics of this cell type were documented a century ago, its potential in development, disease, and wound healing is only now becoming apparent. In the embryo, mesothelia provide vasculogenic cells for the developing heart, lungs, and gut. Furthermore, adult mesothelial cells can be reactivated using thymosin β4 and mobilized to aid in tissue repair. Despite their positive role in development and repair, mesothelia are also susceptible to adhesion and tumor formation. With knowledge that the mesothelium is an important mediator of tissue repair as well as disease, it will be important to identify other factors like thymosin β4 that have the ability to potentiate these cells. Future use of chemical and genetic agents in conjunction with mesothelial cells will lead to enhanced therapeutic potential and mitigation of deleterious outcomes.

Keywords: mesothelium; thymosin β4; smooth muscle regeneration; blood vessel injury

Introduction

In the 1880s, Minot first described the epithelial component of the lateral plate mesoderm that comprises the body cavities of vertebrate organisms.[1] After more than a century of characterization, we have arrived at the term mesothelium to describe the most superficial mesodermally derived layer composed of a simple squamous epithelium that covers all organs contained in the peritoneal, pleural, and pericardial cavities of the adult organisms as well as the cavities themselves. Despite its early identification, it is only recently that a better appreciation of mesothelial development and diversity and its roles in tissue repair as well as disease has come to light. Since this initial description, much progress has been made in our understanding of the origin and function of cardiac mesothelium (the epicardium). Yet, only recently have in-depth studies been devoted to examination of the pleural and peritoneal mesothelia. While similarities exist between these tissues, they are not as Mutsaers suggests, "essentially similar regardless of species or anatom-

ical site."[1] Rather, these distinct tissues have different points of origin as well as different biological potentials. In addition, it is becoming apparent that this normally quiescent lining has the ability to be therapeutic as well as deleterious under certain conditions. Here, we give an overview of mesothelial biology and how thymosin β4 (Tβ4) can be used to potentiate mesothelial-mediated repair.

Mesothelia during development

The development of the mesothelium of the pericardial cavity, the epicardium, has been extensively studied. It is known that these mesothelial cells originate outside of the heart in the proepicardium.[2,3] These cells then migrate over the developing looped heart tube where some of them undergo an epithelial to mesenchymal transformation and dive into the myocardium to contribute smooth muscle[4] and endothelial cells[5] to the coronary vasculature.[6–8] Other cells remain on the surface of the heart to form the epicardium proper, a tissue that remains relatively quiescent in adult organisms.

doi: 10.1111/j.1749-6632.2012.06713.x
Ann. N.Y. Acad. Sci. 1269 (2012) 125–130 © 2012 New York Academy of Sciences.

Recently, it has come to light that this may not be the convention for all types of mesothelium. Gut mesothelium does not originate from a proepicardium-like organ. Rather, a resident population of cells embedded in the splanchnic mesoderm of the developing alimentary canal differentiates into the mesothelium that encompasses the gut (Winters *et al.,* in revision). Some of these cells subsequently undergo an epithelial to mesenchymal transformation and provide the majority of smooth muscle cells for the gut vasculature.[9] The origin of pleural mesothelium is unknown as of yet. However, it has been shown that pleural mesothelial cells are a source of vascular smooth muscle as well as mesenchyme for the developing lungs.[10] These data suggest that while mesothelia may differ on their points of origin, a conserved role for mesothelium in development of coelomic organ vasculature exists.

Functions of adult mesothelia

Historically, mesothelium was though to be the coelom's protective barrier against physical damage and pathogens.[11] Indeed, mesothelial cells secrete glycosaminoglycans, phosphatidylcholine, and surfactant, all of which act as lubricants to reduce friction between coelomic organs and the body wall. In addition, mesothelium is involved in transportation of fluid across coelomic cavities. The luminal aspect of pleural and peritoneal mesothelia is carpeted by microvilli, which increases surface area and promotes fluid absorption.[12,13] Furthermore, stomatal pores are located at mesothelial cell junctions and provide a portal for fluid and cells to be shuttled into the underlying lymphatic system and away from the coelomic space.[14] While these functions were the first to be described, it is evident that mesothelium is actively involved in other aspects of homeostasis and tissue repair.

Mesothelium and disease

Serosal adhesions can form following postoperative injury to the mesothelium. Under normal circumstances, if the mesothelium is damaged, mesothelial cells secrete fibrin in order to initiate tissue repair. Within days, the mesothelial lining is regenerated and the fibrin is reabsorbed. However, if normal regeneration is perturbed, patches of fibrous, vascularized, and innervated tissue form between adjacent organs or between organs and the body wall, which can lead to pain, intestinal obstructions, and infertility.[15]

Cancer of the mesothelium is referred to as mesothelioma. Predominantly found in pleural mesothelium, mesothelioma can also occur at lower frequencies in the peritoneum and pericardial cavity as well.[16] It has been reported that the most common etiological cause of mesothelioma is asbestos exposure. However, tumors can occur without asbestos exposure, suggesting other causative agents. An aggressive form of cancer, mesothelioma is generally unresponsive to conventional chemotherapeutic and radiation therapies.

Mesothelia and wound healing

The healing power of mesothelium has been recognized in clinical circles for well over a century.[17] The omentum, a visceral mesothelial tissue, has been transplanted onto a variety of injured tissues including ischemic hearts,[18] gastric ulcers,[19] and the brains of Alzheimer patients.[20] These procedures demonstrated the omentum's remarkable ability to stimulate new blood vessel formation. In addition, mesothelial cells play a role in modulating coelomic inflammation.[21] Mesothelial secretion of chemokines promotes leukocyte migration into the coelom in response to pathogens or tissue damage.[22] Additionally, in response to injury, mesothelial cells can release growth factors, including transforming growth factor beta (TGF-β) family members, fibroblast growth factors (FGFs), epidermal growth factors (EGFs), and platelet-derived growth factor (PDGF).[23] These growth factors can then regulate aspects of cell proliferation, cell migration, and extracellular matrix deposition in order to promote wound healing.[1] Taken together, these studies illustrate a paracrine role for mesothelium in tissue repair.

Tβ4 and wound healing

Thymosin β4 (Tβ4) was originally identified as a G-actin–sequestering protein [24] and is now known to have a wide range of biological activities, including influencing cell migration. As a regulator of actin cytoskeletal dynamics, Tβ4 can greatly influence the ability of a cell to move.[25] In addition, Tβ4 can activate AKT and PKC signaling cascades, potent inducers of cell migration [26–28] (Fig. 1). Finally, Tβ4

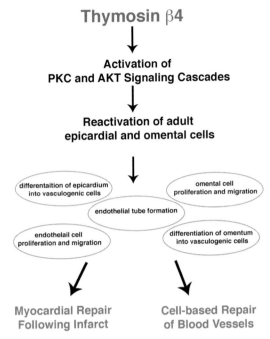

Figure 1. Tβ4 reactivates adult mesothelial cells to aid in tissue repair. Tβ4 can activate PKC and AKT signaling in adult epicardial and omental mesothelial cells. In epicardial cells, Tβ4 promotes proliferation, migration, and differentiation into vasculogenic lineages, and endothelial tube formation, all of which aid in repair of the myocardium following infarct. Similarly, in omental mesothelial cells, Tβ4 promotes proliferation, migration, and differentiation into vasculogenic lineages in order to accelerate the repair of injured blood vessels.

can regulate components of the basement membrane like laminin-5, which is known to affect cell adhesion and migration.[29]

Tβ4 was also identified as a mediator of cell differentiation. Labeled an angiogenic molecule, Tβ4 was shown to induce endothelial and smooth muscle cell differentiation as well as tube formation[27] (Fig. 1). Furthermore, various mouse models demonstrated that Tβ4 promoted keratinocyte and hair follicle differentiation, leading to dermal repair and increased hair growth, respectively.[30,31]

Because of its ability to modulate cell migration and differentiation, Tβ4 was thought to be clinically relevant. Indeed, it has shown promise in promoting corneal[32,33] and dermal[34,35] wound healing. Furthermore, Tβ4 was found to be cardioprotective and can promote cardiac regeneration following infarct.[36–38] Specifically, addition of Tβ4 to a mouse infarct model mobilized adult epicardial cells to in-

vade the myocardium and establish new coronary vessels (Fig. 1).[39]

Potentiating omental grafts with Tβ4

Until recently, the mesothelium was only thought to play a paracrine role in wound healing.[11] And while several studies demonstrated that adult omental mesothelium retained the ability to differentiate into vasculogenic cells,[28,40] we were the first group to demonstrate a cellular contribution of omentum to blood vessel repair.[28] In studies performed in mice, when omentum was grafted onto injured arteries, omental mesothelium differentiated into smooth muscle and remained adjacent to the vessel wall as it healed (Fig. 2A and B). While no graft integration into the injured vessel was observed, the mesothelial cells did provide paracrine signaling that accelerated wound healing.[28] Knowing that Tβ4 can promote cell migration, vessel development, and tissue repair, agarose beads soaked in Tβ4 were added to omental grafts that were transplanted onto injured blood vessels (Fig. 2C). Interestingly, Tβ4 was able to potentiate mesothelial grafts in several ways (Fig. 1). First, it enhanced the ability of omental cells to differentiate into vascular smooth muscle and endothelial cells *in vitro*. In addition, it promoted omental cell migration. Finally, Tβ4 mobilized mesothelial graft cells to integrate into the wounded vessel and repopulate the smooth muscle layers of the vessel thereby significantly decreasing healing time of grafted vessels *in vivo*. Furthermore, myographic studies revealed that vessels that incorporated mesothelial-derived smooth muscle cells were functionally indistinguishable from uninjured vessels.[28] This was the first indication that omentum was capable of cellular contributions to healing tissues. In addition, it was the first report that factors like Tβ4 could be used to potentiate omental grafts and improve therapeutic outcomes.

Conclusions and future directions

Coelomic epithelia was identified well over 100 years ago, however, we have only recently begun to appreciate the impact mesothelia can have on coelomic organ development, repair, and disease. As new data emerges, it is becoming clear that all mesothelium is not created equal. While most studies have concentrated on characterizing the origin, developmental contribution, and wound-healing potential of the

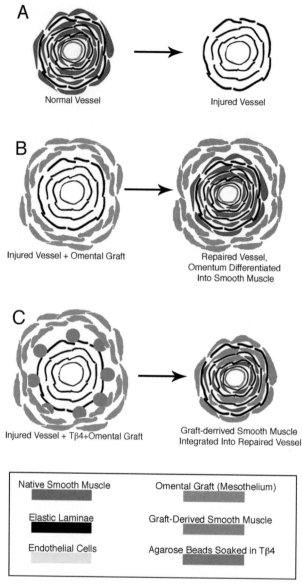

Figure 2. Thymosin β4 mobilizes omental cells for blood vessel repair. Murine carotid arteries were subjected to cryo-injury, rendering them devoid of vascular smooth muscle and endothelial cells (A). When omental mesothelial grafts were transplanted onto injured arteries, the grafts differentiated into smooth muscle and accelerated vessel healing via a paracrine mechanism. However, no integration of mesothelial cells into healing vessels was observed (B). Agarose beads soaked in Tβ4 were transplanted onto injured arteries in conjunction with omental grafts. Tβ4 mobilized omental-derived smooth muscle cells and promoted their incorporation into the medial layers of vessels as they healed, thereby significantly accelerating the healing process (C).

epicardium, the nuances of pleural and peritoneal mesothelia are less well defined. A recent report suggests that at least two types of mesothelial tissues are derived from different source tissue (Winters *et al.*, in revision), and yet, despite variations in their developmental origins, all mesothelium studied to date is able to provide vasculogenic cells to the or-gans it surrounds.[4,9,10] Similarly, the ability to reactivate adult epicardial and omental mesothelium has been established.[28,39] And in both instances, Tβ4 is able to enhance the ability of each tissue to facilitate wound healing by promoting cell mobilization and differentiation (Fig. 1). Understanding how to potentiate mesothelial cells in order to maximize their therapeutic potential will be important for future studies. To this end, advances in the field of small organic molecule development will be instrumental in harnessing mesothelial potential. Likewise, as not all migration and differentiation associated with mesothelial cells is beneficial, understanding the mechanisms of mesothelial disease will also be of broad interest to the clinical and basic science communities.

Conflicts of interest

The authors declare no conflicts of interest.

References

1. Mutsaers, S.E. 2004. The mesothelial cell. *Int. J. Biochem. Cell. Biol.* **36:** 9–16.
2. Viragh, S. *et al.* 1993. Early development of quail heart epicardium and associated vascular and glandular structures. *Anat. Embryol.* **188:** 381–393.
3. Viragh, S. & C.E. Challice. 1981. The origin of the epicardium and the embryonic myocardial circulation in the mouse. *Anat. Rec.* **201:** 157–168.
4. Mikawa, T. & R.G. Gourdie. 1996. Pericardial mesoderm generates a population of coronary smooth muscle cells migrating into the heart along with ingrowth of the epicardial organ. *Dev. Biol.* **174:** 221–232.
5. Perez-Pomares, J.M. *et al.* 2002. Origin of coronary endothelial cells from epicardial mesothelium in avian embryos. *Int. J. Dev. Biol.* **46:** 1005–1013.
6. Dettman, R.W. *et al.* 1998. Common epicardial origin of coronary vascular smooth muscle, perivascular fibroblasts, and intermyocardial fibroblasts in the avian heart. *Dev. Biol.* **193:** 169–181.
7. Manner, J. *et al.* 2001. The origin, formation and developmental significance of the epicardium: a review. *Cells Tissues Organs* **169:** 89–103.
8. Manner, J. 1993. Experimental study on the formation of the epicardium in chick embryos. *Anat. Embryol.* **187:** 281–289.
9. Wilm, B. *et al.* 2005. The serosal mesothelium is a major source of smooth muscle cells of the gut vasculature. *Development* **132:** 5317–5328.
10. Que, J. *et al.* 2008. Mesothelium contributes to vascular smooth muscle and mesenchyme during lung development. *Proc. Natl. Acad. Sci. USA* **105:** 16626–16630.
11. Mutsaers, S.E. 2002. Mesothelial cells: their structure, function and role in serosal repair. *Respirology* **7:** 171–191.
12. Andrews, P.M. & K.R. Porter. 1973. The ultrastructural morphology and possible functional significance of mesothelial microvilli. *Anat. Rec.* **177:** 409–426.
13. Tsilibary, E.C. & S.L. Wissig. 1977. Absorption from the peritoneal cavity: SEM study of the mesothelium covering the peritoneal surface of the muscular portion of the diaphragm. *Am. J. Anat.* **149:** 127–133.
14. Ohtani, O., Y. Ohtani & R.X. Li. 2001. Phylogeny and ontogeny of the lymphatic stomata connecting the pleural and peritoneal cavities with the lymphatic system–a review. *Ital. J. Anat. Embryol.* **106:** 251–259.
15. Herrick, S.E. *et al.* 2000. Human peritoneal adhesions are highly cellular, innervated, and vascularized. *J. Pathol.* **192:** 67–72.
16. Moore, A.J., R.J. Parker & J. Wiggins. 2008. Malignant mesothelioma. *Orphanet. J. Rare. Dis.* **3:** 34.
17. Cannaday, J.E. 1948. Some uses of undetached omentum in surgery. *Am. J. Surg.* **76:** 502–505.
18. Taheri, S.A. *et al.* 2005. Myoangiogenesis after cell patch cardiomyoplasty and omentopexy in a patient with ischemic cardiomyopathy. *Tex. Heart Inst. J.* **32:** 598–601.
19. Matoba, Y., H. Katayama & H. Ohami. 1996. Evaluation of omental implantation for perforated gastric ulcer therapy: findings in a rat model. *J. Gastroenterol.* **31:** 777–784.
20. Goldsmith, H.S. 2007. Omental transposition in treatment of Alzheimer disease. *J. Am. Coll. Surg.* **205:** 800–804.
21. Litbarg, N.O. *et al.* 2007. Activated omentum becomes rich in factors that promote healing and tissue regeneration. *Cell Tissue Res.* **328:** 487–497.
22. Visser, C.E. *et al.* 1998. Chemokines produced by mesothelial cells: huGRO-alpha, IP-10, MCP-1 and RANTES. *Clin. Exp. Immunol.* **112:** 270–275.
23. Mutsaers, S.E. *et al.* 1997. Mechanisms of tissue repair: from wound healing to fibrosis. *Int. J. Biochem. Cell Biol.* **29:** 5–17.
24. Safer, D., M. Elzinga & V.T. Nachmias. 1991. Thymosin beta 4 and Fx, an actin-sequestering peptide, are indistinguishable. *J. Biol. Chem.* **266:** 4029–4032.
25. Crockford, D. *et al.* 2010. Thymosin beta4: structure, function, and biological properties supporting current and future clinical applications. *Ann. N.Y. Acad. Sci.* **1194:** 179–189.
26. Bock-Marquette, I. *et al.* 2004. Thymosin beta4 activates integrin-linked kinase and promotes cardiac cell migration, survival and cardiac repair. *Nature* **432:** 466–472.
27. Bock-Marquette, I. *et al.* 2009. Thymosin beta4 mediated PKC activation is essential to initiate the embryonic coronary developmental program and epicardial progenitor cell activation in adult mice *in vivo*. *J. Mol. Cell Cardiol.* **46:** 728–738.
28. Shelton, E.L. *et al.* 2012. Omental grafting: a cell-based therapy for blood vessel repair. *J. Tissue Eng. Regen. Med.* doi: 10.1002/term.528
29. Sosne, G. *et al.* 2004. Thymosin beta 4 stimulates laminin-5 production independent of TGF-beta. *Exp. Cell Res.* **293:** 175–183.
30. Francis Godschalk, M. 2007. Pressure ulcers: a role for thymosin beta4. *Ann. N.Y. Acad. Sci.* **1112:** 413–417.
31. Philp, D. *et al.* 2004. Thymosin beta4 increases hair growth by activation of hair follicle stem cells. *FASEB J.* **18:** 385–387.
32. Sosne, G. *et al.* 2002. Thymosin beta 4 promotes corneal wound healing and decreases inflammation *in vivo* following alkali injury. *Exp. Eye Res.* **74:** 293–299.
33. Sosne, G. *et al.* 2001. Thymosin beta 4 promotes corneal wound healing and modulates inflammatory mediators *in vivo*. *Exp. Eye Res.* **72:** 605–608.
34. Philp, D. *et al.* 2003. Thymosin beta 4 and a synthetic peptide containing its actin-binding domain promote dermal wound repair in db/db diabetic mice and in aged mice. *Wound Repair Regen.* **11:** 19–24.

35. Fine, J.D. 2007. Epidermolysis bullosa: a genetic disease of altered cell adhesion and wound healing, and the possible clinical utility of topically applied thymosin beta4. *Ann. N.Y. Acad. Sci.* **1112:** 396–406.
36. Marx, J. 2007. Biomedicine. Thymosins: clinical promise after a decades-long search. *Science* **316:** 682–683.
37. Hinkel, R. *et al.* 2008. Thymosin beta4 is an essential paracrine factor of embryonic endothelial progenitor cell-mediated cardioprotection. *Circulation* **117:** 2232–2240.
38. Smart, N. *et al.* 2007. Thymosin beta4 induces adult epicardial progenitor mobilization and neovascularization. *Nature* **445:** 177–182.
39. Smart, N. *et al.* 2010. Thymosin beta4 facilitates epicardial neovascularization of the injured adult heart. *Ann. N.Y. Acad. Sci.* **1194:** 97–104.
40. Kawaguchi, M., D.M. Bader & B. Wilm. 2007. Serosal mesothelium retains vasculogenic potential. *Dev. Dyn.* **236:** 2973–2979.

Ann. N.Y. Acad. Sci. ISSN 0077-8923

ANNALS OF THE NEW YORK ACADEMY OF SCIENCES
Issue: *Thymosins in Health and Disease*

Originally published as: Low, T.L.K. Current status of thymosin research: evidence for the existence of a family of thymic factors that control T-cell maturation. *Ann. N.Y. Acad. Sci.* **332**: 33–48.

CURRENT STATUS OF THYMOSIN RESEARCH: EVIDENCE FOR THE EXISTENCE OF A FAMILY OF THYMIC FACTORS THAT CONTROL T-CELL MATURATION *

Teresa L. K. Low, Gary B. Thurman, Carolina Chincarini,
John E. McClure, Gailen D. Marshall, Shu-Kuang Hu,
and Allan L. Goldstein

Department of Biochemistry
George Washington University School of Medicine and
Health Sciences
Washington, D.C. 20037

INTRODUCTION

In 1974 at the last New York Academy of Sciences Symposium on "Thymus Factors in Immunity," we presented data indicating that the varied biological activities ascribed to the partially purified thymosin fraction 5 may not reside within a single molecular species but rather with multiple thymic factors acting individually, sequentially, or in concert to endow the host with its normal complement of immunity. At that time, we had purified from thymosin fraction 5 a protein (termed fraction 8) with a molecular weight of approximately 12,000 daltons. Fraction 8 could be further dissociated by gel-filtration on Sephadex G-50 equilibrated with 6M guanidinium-C1 into two smaller subunits of approximately 3,200 and 3,400. Although fraction 8 was more active than fraction 5 in the induction of surface markers (TL and thy-1), it was less active in other assays such as in mixed lymphocyte reaction (MLR) and murine rosette bioassay.[1] We have now confirmed and extended these preliminary observations and have purified to homogeneity over 16 of the individual peptide components of fraction 5. Several of these components have been chemically characterized including sequence analysis of thymosins $\alpha_1{}^{2-4}$, β_3 and β_4 (in preparation) and a ubiquitous peptide termed $\beta_1{}^{3-6}$. Ongoing studies suggest that some of these peptides, as illustrated in FIGURE 1, may act at different sites and on different subsets of T-cells and contribute to the normal maintenance of immune function and balance. In this paper, we report the current status of the chemical characterization and biological activity of these peptides. We will also summarize the current status of the clinical trials of thymosin fraction 5 in patients with primary immunodeficiency diseases,[7–10] cancer,[10–15] and autoimmune diseases.[16]

BIOCHEMISTRY OF THYMOSIN

The identification of a lymphocytopoietic factor in rat and mouse thymus extracts was first described in 1965 by Klein, Goldstein, and White.[17, 18] In

* This research is supported in part by grants from the National Cancer Institute (CA 16964 and CA 20667), Hoffmann-La Roche, Inc., and the John A. Hartford Foundation, Inc.

doi: 10.1111/j.1749-6632.2012.06765.x

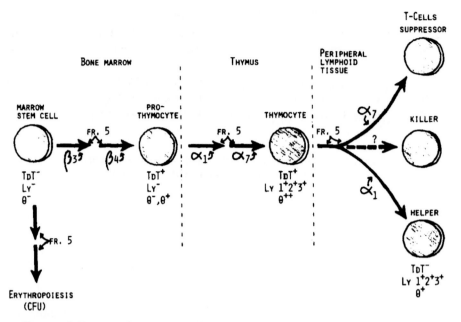

FIGURE 1. Proposed sites of action of thymosin polypeptides on maturation of T-cell subpopulations.

1966, a stable form (acetone powder) of the biologically active fraction was prepared from calf thymus and was termed thymosin fraction 3.[19] After further fractionation using a combination of ion-exchange chromatographic procedures and/or preparative polyacrylamide gel electrophoresis, two more purified fractions, termed thymosin fraction 7 [20] and thymosin fraction 8 [1] were prepared. In 1975 we reported the modification of the early fractionation procedures to allow for the preparation of larger quantities of the biologically active thymosin fraction for projected clinical trials.[1] This preparation was termed thymosin fraction 5.

Purification and Chemical Characterization of Thymosin Fraction 5

As shown in FIGURE 2, thymosin fraction 5 is prepared from calf thymus as described by Hooper *et al.*[1] In addition, the 50–95% ammonium sulfate precipitation cut was also collected and termed thymosin fraction 5A.[3] Thymosin fraction 5 has been demonstrated to be an effective immunopotentiating agent and can act in lieu of the thymus gland to reconstitute immune function in certain thymic-deprived or immunodeprived individuals.[21] Fraction 5 is a partially purified preparation and is composed of a group of polypeptides that are heat stable up to 80° C and with molecular weight ranging between 1,000 and 15,000. The preparation contains small amounts of carbohydrate and is essentially free of nucleotides and lipids.

On an isoelectric focusing gel, fraction 5 is shown to consist of 30 or more individual components. Since there are now many laboratories working with

Ann. N.Y. Acad. Sci. 1269 (2012) 131–146 © 2011 New York Academy of Sciences.

thymosin and other thymic factors, we have proposed a nomenclature system based on this isoelectric focusing pattern to facilitate identification and comparison of thymic peptides from one laboratory to another.[2] As shown in FIGURE 3, the separated polypeptides are divided into three regions. The α region consists of polypeptides with isoelectric points below 5.0; the β region, 5.0 to 7.0; and the γ region, above 7.0. The subscript numbers α_1, α_2, β_1, β_2, etc. are used to identify the polypeptides from each region in the order they are isolated.

Fractionation, Purification, and Chemical Characterization of Thymosin Polypeptides

Our major and continuing efforts have been directed toward the further fractionation and isolation of the biologically active components in thymosin fraction 5. Using a combination of ion-exchange chromatography and gel-filtration, monitored by analytical isoelectric focusing in polyacrylamide gels, we have isolated 16 (α_1, to α_{10}, and β_1 to β_6) polypeptide components from fraction 5 for further characterization.

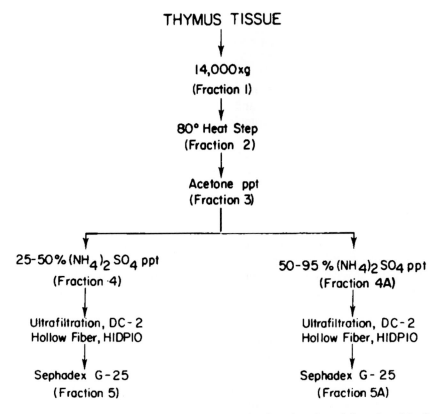

FIGURE 2. Purification of bovine thymosin fraction 5 and fraction 5A. One kilogram of thymus tissue homogenized in 3L of NaCl provides the sample to initiate the procedure (from 2).

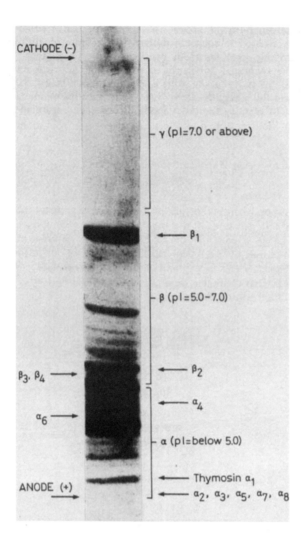

CATHODE (−)

γ (pI=7.0 or above)

β₁

β (pI=5.0–7.0)

β₃, β₄

β₂

α₄

α₆

α (pI=below 5.0)

Thymosin α₁

ANODE (+)

α₂, α₃, α₅, α₇, α₈

FIGURE 3. Isoelectrically focused sample of thymosin fraction 5 showing distribution of mostly acidic peptide components. The α, β, and γ regions indicate segmentation of the gel based upon isoelectric point (pI) ranges suggested for peptide nomenclature to facilitate interlaboratory comparison of purified thymic extracts (from 2).

Thymosin α_1

This is the first thyrnosin polypeptide to be purified to homogeneity.[2] Thymosin α_1 is very active in several bioassay systems, including the human E-rosette assay, guinea pig MIF assay, and murine Lyt[1+, 2+, 3+] induction and helper cell assays (A. Ahmed, these proceedings). The purification procedure has been described [2, 3] and is diagrammed in FIGURE 4. Thymosin α_1 migrates as a single band on analytical gels at pH 8.3 and 2.9 and as major band with an isoelectric point of 4.2 on an isoelectric focusing slab gel of pH range 3–5. The yield of thymosin α_1 from fraction 5 is about 0.6% and is free of carbohydrate, lipid, or nucleotides. The amino acid sequence of

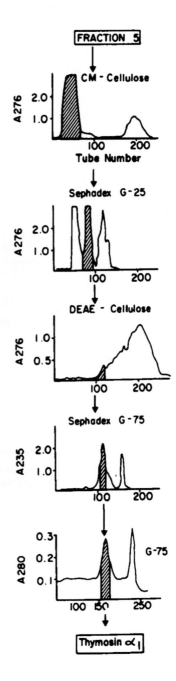

FIGURE 4. Flow diagram of the fractionation of thymosin α_1 from bovine thymosin fraction 5. Tube numbers are shown on the abscissa, absorbance on the ordinate of each plot. The shaded areas indicate pooled fractions collected for use in the subsequent step. Material from the final step represents a yield of 0.6% and migrates as a single band on analytical PA gels at pH 8.3 or 2.9 and as a major band with pI of 4.2 on isoelectric focusing slab gel, pH 3–5 (from 3).

thymosin α_1 has been determined and the detailed sequencing data of this molecule have been reported.[4] Thymosin α_1 is composed of 28 amino acid residues with a molecular weight of 3,108. As shown in FIGURE 5, the amino terminus of thymosin α_1 is blocked by an acetyl group. There are six pairs of repeating amino acid residues along the sequence of thymosin α_1. These residues are shaded in FIGURE 5. However, the significance of these observations is not clear.

Comparison of the amino acid sequence of thymosin α_1 with the published sequence of other thymic factors such as thymopoietin [23] and facteur thymique sérique [24] reveals no homology. Thymosin α_1 has been chemically synthesized by Wang, Kulesha, and Winter.[25] The synthetic α_1 has been tested in MIF and E-rosette assays and appears to have similar activity to the natural thymosin α_1. Most recently, Freire and coworkers [26] have carried out translation and *in vitro* synthesis of a precursor of thymosin α_1 using messenger RNA from calf thymus in a cell-free wheat-germ system. The radioactive products that were immunoprecipitable with antisera against thymosin fractions were analyzed to be identical to those expected for tryptic peptides from thymosin α_1. These studies indicate that thymosin α_1 is produced in the thymus and is probably synthesized as a longer peptide chain of 16,000 daltons.

As summarized in TABLE 1, thymosin α_1 is 10 to 1,000 times as active as thymosin fraction 5 in certain functional assays but is not active in others.

Thymosin α_7

This partially purified preparation is prepared from thymosin fraction 5 by ion-exchange chromatography on CM-cellulose and DEAE-cellulose and

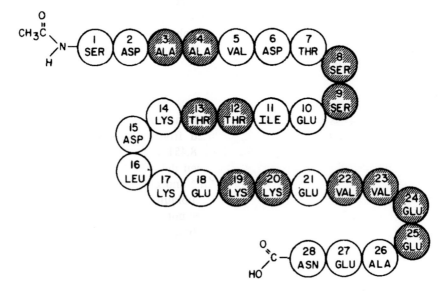

FIGURE 5. Complete amino acid sequence of bovine thymosin α_1. The shaded areas are the repeating amino acid residues along the polypeptide chain.

TABLE 1

SUMMARY OF THYMOSIN ACTIVITY IN VARIOUS BIOASSAYS

Assay	Thymosin Fraction 5 (μg)	Polypeptide β_1 (Ubiquitin)	Thymosin α_1 (μg)
MLR	1–10	N.A.*	N.A.
MIF	1–5	N.A.	0.001–0.1
E-rosette	1–10	N.A.	0.001–0.01
Mitogen	1–10	N.A.	0.01–0.1
Antibody	10–100	N.A.	N.A.
Lyt [1+, 2+, 3+]	10–100	N.A.	0.01–0.05
TdT	0.01–0.1,† 100‡	N.A.	N.A.†

* N.A.=Not active.

† Induction of TdT in bovine serum albumin separated bone marrow cells *in vitro* from athymic nu/nu mice.

‡ Induction of TdT *in vivo* in thymocytes from hydrocortisone acetate suppressed C57BL/6J mice.

gel-filtration on Sephadex G-75 (Low and Goldstein, in preparation). It is highly acidic with an isoelectric point around 3.5 and a molecular weight of 2,200. The amino acid composition of this molecule revealed that this molecule does not contain any basic amino acid residues, but contains about 50% of Asx (aspartic acid or asparagine) and Glx (glutamic acid or glutamine). A. Ahmed *et al.* have observed that thymosin α_7 induces Lyt [1+, 2+, 3+] cells in the B layer of bovine serum albumin-gradient separated bone marrow cells.[22] In addition, S. Horowitz (personal communication) has found that thymosin α_7 reconstitutes suppressor cell function *in vitro* in lymphocytes of patients with systemic lupus erythematosus (SLE) . Elsewhere in these proceedings, Ahmed *et al.* have confirmed the observation that α_7 induces suppressor cells *in vitro*. These results suggest thymosin α_7 is acting on prothymocytes to form cells with suppressor function.

Polypeptide β_1

Polypeptide β_1 is the predominant band on the isoelectric focusing gel of fraction 5 (see FIGURE 3). It was isolated from thymosin fraction 5A as described.[3] It has a molecular weight of 8,451 and an isoelectric point of 6.7. The amino acid sequence of β_1 has been determined [4] and has revealed that this molecule is identical to ubiquitin [5] and a portion of protein A24, a nuclear chromosomal protein.[27] Recently, ubiquitin was also isolated from trout testis chromatin.[28] Polypeptide β_1 does not show biological activity (see TABLE 1) in our bioassay systems, indicating that it is not an important molecule for T-cell maturation.

Thymosin β_1 and β_4

Both partially purified thymosin β_3 and β_4 preparations were prepared from thymosin fraction 5A by DEAE-cellulose and gel-filtration on Sephadex

G-75 (Low and Goldstein, in preparation). Thymosin β_3 has an isoelectric point of 5.2 and a molecular weight of approximately 5,500. The thymosin β_4 has an isoelectric point of 5.1 and a molecular weight of approximately 5,250. From the partial sequence obtained from these polypeptides (Low and Goldstein, unpublished data), they appear to share an identical sequence through most of their amino-terminal part (about 45 residues) and differ in the carboxyl-terminal ends (about 8 residues). Both preparations induce terminal deoxynucleotidyl transferase (TdT) positive cells *in vitro* in the A and B layers of bovine serum albumin-separated bone marrow cells of nude mice.[29] They also accelerate the repopulation of TdT-positive cells *in vivo* in thymocytes of hydrocortisone acetate-treated C57B1/6J mice (Hu, Low and Goldstein, in preparation). From these results, it has been postulated that thymosin β_3 and β_4 may be acting on stem cells to form prothymocytes.

Thymosin Fraction 5 and α_1 From Other Species

In order to evaluate the species variation of thymosin polypeptides, we prepared thymosin fraction 5 from several different species including human, pig, sheep and chinchilla thymuses. The human thymus tissue was obtained from tissue excised during open heart surgery and from selected autopsies. Thymosin α_1, from several animal species, was prepared from fraction 5 using a modification of the extraction and fractionation procedures developed from the isolation of bovine αt. Human, porcine and ovine thymosin α_1 have been partially sequenced. From the results obtained, they appear to have an identical sequence to bovine α_1.

ENDOCRINE ROLE OF THE THYMUS GLAND

A prerequisite for establishing endocrine function for any organ is the demonstration that cell-free extracts of the tissue will replace, in whole or in part, the specific biological functions that have been assigned to that organ. This has now been adequately demonstrated for the thymus. Reconstitution studies with crude and partially purified thymic extracts from various animal sources—mouse, calf, pig, rat, sheep, guinea pig, and man—have established that cell-free preparations can act in lieu of an intact thymus and restore many of the deficiencies due to removal or dysfunction of the gland.[30] Ongoing studies point to a major role for thymosin and other thymic factors in maintaining immune balance (FIGURE 6). There is now a considerable body of evidence indicating that genetic, viral, radiation, or chemical insults to the thymus may result in severe immune imbalances involving both T- and B-cells and may contribute to a number of serious diseases.[30]

Activity similar to that of thymosin has been demonstrated in the blood of normal adult mice by means of a rosette bioassay and appears in the blood of immunodeficient patients shortly after administration of thymosin. The development of rabbit antibodies to bovine thymosin polypeptides has provided confirmatory evidence using RIA and hemagglutinin techniques that circulating thymosin is detectable in the blood of all mammals tested (John E. McClure and Allan L. Goldstein, in preparation). We have also found that levels of circulating thymosin decrease with age in man and other mammals, and ongoing studies indicate that thymosin levels are altered in patients with specific immunodeficiency diseases and some malignancies (unpublished observations). Such observations fulfill a second requirement for designating an organ as endocrine in function, i.e., its secretory products can be demonstrated in the circulation. These results suggest that the thymus, in addition to influencing endogenous lymphoid stem cells and cells traversing the gland *in situ*, may affect the behavior of certain populations of immature lymphoid cells outside the thymus.

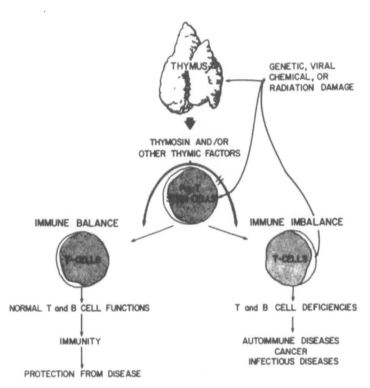

FIGURE 6. Thymus contribution to immunity. Impairment of lymphoid elements including the thymus and immunocompetent cells, by various deleterious agents causes deficiencies in immunity which may lead to a variety of disease manifestations. These conditions may in part be the result of a malfunctioning thymus and inadequate levels of thymic secretions.

REVIEW OF BIOLOGICAL STUDIES WITH THYMOSIN

Our own efforts initiated in the laboratory of Dr. Abraham White at the Albert Einstein College of Medicine in New York have contributed to the elucidation of the endocrine function of the thymus gland in the development and maintenance of the immune system.[11, 14, 17, 18, 31–38] In collaboration with our clinical colleagues, we were the first to successfully initiate clinical trials with thymosin in patients with primary immunodeficiency diseases [14, 39, 40] and advanced malignancies.[10, 11, 41–47]

Studies in Animals

Thymosin preparations have been found to be effective in partially or fully inducing and maintaining immune function in a variety of normal and immuno-deficient animal models.[1–4, 7, 14, 17–20, 28, 31–37, 48–56] Thymosin treatment has been shown to increase the survival of neonatally thymectomized mice [35, 36] accelerate skin graft rejection,[32, 33] and restore or accelerate graft-versus-host reactions[31] and development of immune functions in newborn mice.[35, 37, 53, 54] Thymosin affects the responsivity of lymphocytes from nude,[57–59, 78] normal,[58–61] and tumor-bearing mice,[62–64] and causes immature mouse lymphoid cells to acquire distinctive T-cell surface antigens.[37, 65] Recent experimental approaches utilizing thymosin have shown its effectiveness in modulating immune parameters and inducing specific types of lymphocytes (killer, helper, and suppressor cells),[59, 65–67] and have shown that specific purified thymosin peptides isolated from thymosin fraction 5 can induce certain markers (TdT, thy-1, and Lyt) and functional expressions of lymphocyte maturation.[52, 59, 77]

There have also been a number of interesting reports that point to a potential role for thymosin in the aging process and in the treatment of autoimmune diseases. These studies have shown that:

—Thymosin treatment of NZB mice can reconstitute suppressor cell and other T-cell functions and temporarily cause remission of some of the symptoms of autoimmune disease commonly seen in these mice, including induction of suppressor T-cells and restoration of antigen-induced depression of DNA synthesis.[59–61]

—Administration of thymosin significantly increases the hemagglutinin response of aged mice.[62]

—*In vitro* administration of thymosin to aging thymectomized rats shortens the median $S + G_2$ phase of the cell cycle to levels seen in young rats.[75]

Studies in Humans

In Vitro

Incubation with thymosin *in vitro* increases the percent of total E-rosette levels of some patients with primary immunodeficiencies,[7] cancer,[14, 48, 74, 79] allergies (J. Hobbs, personal communication), severe burns,[80] leprosy (unpublished observations), viral infection,[81] SLE, rheumatoid arthrities,[82, 83] and in aged normal individuals[84] (S. Ishizawa, personal communication).

In a study of 388 patients with head and neck, medastinal, and pelvic malignancies during radiation therapy, and of 277 normal adults, the *in vitro* positive response to thymosin correlated with radiation portal for the cancer patients, and with initial T-cell levels for both groups.[79]

Incubation with thymosin *in vitro* induces the appearance of suppressor T-cells in the peripheral lymphocytes of patients with both active and inactive

SLE as measured in a MLR assay and can induce the expression of increased immunoglubulin production assays.[85]

Incubation with thymosin of a subpopulation of null cells isolated by bovine serum albumin density gradient from normal adult controls results in the induction of T-cell rosettes. These same cells have enhanced responsiveness to phytohemagglutinin and have increased MLC capacities.[86] Within the null cell compartment, following incubation with thymosin fraction 5, no changes in B-cells, monocytes, Fc-positive or C3-receptor-bearing cells could be documented. Therefore, in normal adult controls, putative stem cells can be induced to form T-cell rosettes following incubation with thymosin fraction 5.

Cells contained within the null cell HTLA⁺ compartment enriched by sequential nylon column filtration and E-rosette depletion could be induced by thymosin fraction 5 incubation *in vitro* to form T-cell rosettes; HBLA⁺ cells were not altered by thymosin incubation.[87] Thus, a subpopulation of null cells isolated from normal peripheral blood lymphocytes can be induced to form T-cell rosettes by incubation with thymosin fraction 5.

TABLE 2

In Vivo INDUCTION OF TdT IN THYMOCYTES
FROM HYDROCORTISONE ACETATE-TREATED C57BL/6J MICE

Treatment	TdT Specific Activity *
Control saline	836.3
Spleen fraction 5 (100 μg/injection)	908.5
Thymosin fraction 5 (100 μg/injection)	2679.8
Thymosin β_3 (1 μg/injection)	2758.6
Thymosin β_4 (10 μg/injection)	2954.3

* Values in p moles ^{8}HdGTP/30 min/10^8 cells.

In Vivo *TdT Assay for Thymosin*

An *in vivo* TdT assay was recently developed in our laboratory for evaluating thymosin activity. Different doses of thymosin fraction 5 and other purified thymosin polypeptides are injected into hydrocortisone acetate-treated C57BL/6J mice daily for 9 to 11 days. The animals are sacrificed, thymocytes prepared, and TdT activity determined using the method described by Pazmino *et al.*[88] The ability of thymosin to induce the differentiation of pre-T cells to TdT-positive thymocytes is demonstrated by the increase in TdT activity. TABLE 2 gives the results of one typical assay which indicates that 1 μg of β_3 per injection is as effective *in vivo* as 100 μg of thymosin fraction 5.

CLINICAL TRIALS WITH THYMOSIN

Primary Immunodeficiency Diseases

Over 50 children have received thymosin for a variety of primary immunodeficiency diseases.[7, 8, 10, 39, 89] These patients have been treated with injections of thymosin up to 400 mg/M² for periods of over four and one-half years (usually daily for two to four weeks, then once per week). Most of the patients have received 60 mg/M² thymosin by subcutaneous injection. To date, there has been no evidence of central nervous system (CNS), liver, kidney, or bone marrow toxicity in this group due to thymosin administration. More than 80% of the pediatric patients who have responded *in vitro* in the E-rosette assay have also responded *in vivo*. A significant increase in T-lymphocyte number and function has been observed in over 40% of the patients studied who responded *in vitro* (E-rosette, MLR), as well as significant clinical improvement in a number of cases, particularly the children with thymic hypoplasias, DiGeorge Syndrome, and Wiskott-Aldrich Syndrome.[7, 8, 47] *In vivo* treatment of patients with thymosin has also been found to induce appearance of serum thymic factor.[90]

Phase I and Phase II Cancer Trials

More than 150 cancer patients have been treated according to Phase I[10, 11, 41, 47] or Phase II protocols.[42–45] Cancer patients have been treated for periods of up to four years. As with pediatric patients, no major side effects have been seen in the majority of patients.

The first Phase II randomized trial of thymosin has now been completed in nonresectable small cell carcinoma of the lungs by Dr. Paul D. Chretien (NCI) and Dr. Martin Cohen and associates at the Washington V.A. Hospital.[42–45] Details of this study are presented elsewhere in these proceedings (see Chretien *et al.*). In this trial, thymosin fraction 5 was found to prolong the survival of cancer patients significantly when given in conjunction with intensive chemotherapy. Mean survival time was increased from 240 days with chemotherapy alone to over 450 days with chemotherapy plus 60 mg/M² thymosin twice per week for the first six weeks of the chemotherapy induction period.

Six of the original 21 patients in the high-dose thymosin group are still alive and tumor-free at over two years (M. Cohen, personal communication).

Autoimmune Disease

To date (see Costanzi *et al.*, these proceedings), six patients with autoimmune diseases have been treated with thymosin fraction 5 in Phase I study for periods ranging from 4 to 16 months.[16] Five of the patients had systemic lupus erythematosus (SLE) and the sixth rheumatoid arthritis. Thymosin administration resulted in improvement in immunological parameters and a major decrease in a cytotoxic serum factor that is present in the sera of many patients with autoimmune diseases. This heterologous factor causes the lysis of murine thymocytes in the presence of complement. Based on these encouraging findings, Phase II randomized trials are planned to determine the efficacy of thymosin therapy in SLE. Although the mechanism of immune reconstitution with thymosin fraction 5 in persons with autoimmune disease is not as yet defined, it may be related in part to induction of a subpopulation of thymosin-activated suppressor or regulator cells.

SUMMARY

Thymosin fraction 5 contains several distinct hormonal-like factors which are effective in partially or fully inducing and maintaining immune function. Several of the peptide components of fraction 5 have been purified, sequenced and studied in assay systems designed to measure T-cell differentiation and function. These studies indicate that a number of the purified peptides act on different subpopulations of T-cells (see FIGURE 1). Thymosin β_3 and β_4 peptides act on terminal deoxynucleotidyl transferase (TdT) negative precursor T-cells to induce TdT positive cells. Thymosin α_1 induces the formation of functional helper cells and conversion of Lyt$^-$ cells to Lyt$^{1+, 2+, 3+}$ cells. Thymosin α_7 induces the formation of functional suppressor T-cells and also converts Lyt$^-$ cells to Lyt$^{1+, 2+, 3+}$ cells. These studies have provided further evidence that the thymus secretes a family of distinct peptides which act at various sites of the maturation sequence of T-cells to induce and maintain immune function.

Phase I and Phase II clinical studies with thymosin in the treatment of primary immunodeficiency diseases, autoimmune diseases, and cancer point to a major role of the endocrine thymus in the maintenance of immune balance and in the treatment of diseases characterized by thymic malfunction.

It is becoming increasingly clear that immunological maturation is a process involving a complex number of steps and that a single factor initiating a single cellular event might not be reflected in any meaningful immune reconstitution unless it is the only peptide lacking.

Given the complexity of the maturation sequence of T-cells and the increasing numbers of T-cell subpopulations that are being identified, it would be surprising if a single thymic factor could control all of the steps and populations involved. Rather, it would appear that the control of T-cell maturation and function involves a complex number of thymic-specific factors and other molecules that rigidly control the intermediary steps in the differentiation process.

REFERENCES

1. HOOPER, J. A., M. C. McDANIEL, G. B. THURMAN, G. H. COHEN, R. S. SCHULOF & A. L. GOLDSTEIN. 1975. Ann. N. Y. Acad. Sci. **249**: 125.
2. GOLDSTEIN, A. L., T. L. K. LOW, M. McADOO, J. McCLURE, G. B. THURMAN, J. L. ROSSIO, C. Y. LAI, D. CHANG, S. S. WANG, C. HARVEY, A. H. RAMEL & J. MEIENHOFER. 1977. Proc. Nat. Acad. Sci. USA **74**: 725.
3. LOW, T. L. K., G. B. THURMAN, M. McADOO, J. McCLURE, J. L. ROSSIO, P. H. NAYLOR & A. L. GOLDSTEIN. 1979. J. Biol. Chem. **254**: 981.
4. LOW, T. L. K. & A. L. GOLDSTEIN. 1979. J. Biol. Chem. **254**: 987.
5. SCHLESINGER, D. H., G. GOLDSTEIN & H. D. NIALL. 1975. Biochemistry **14**: 2214–2218.
6. GOLDSTEIN, G., M. SCHEID, U. HAMMERLING, E. A. BOYSE, D. H. SCHLESINGER & H. D. NIALL. 1975. Proc. Nat. Acad. Sci. USA **72**: 11–15.
7. WARA, D. W., A. L. GOLDSTEIN, W. DOYLE & A. J. AMMANN. 1975. N. Engl. J. Med. **292**: 70.
8. WARA, D. & A. AMMANN. 1978. Transplant. Proc. **10**: 11.
9. GOLDSTEIN, A. L., D. W. WARA, A. J. AMMANN, H. SAKAI, N. S. HARRIS, G. B. THURMAN, J. A. HOOPER, G. H. COHEN, A. S. GOLDMAN, J. J. COSTANZI & M. C. McDANIEL. 1975. Transplant. Roc. 7: 681.

10. ROSSRO, J. L. & A. L. GOLDSTEIN. 1977. World J. Surg. **1**: 605.
11. COSTANZI, J. J., R. G. GAGLIANO, D. LOUKAS, F. DELANEY, H. SAKAI, N. S. HARRIS, G. B. THURMAN & A. L. GOLDSTEIN. 1977. Cancer **40**: 14.
12. GOLDSTEIN, A. L., G. B. THURMAN, J. L. ROSSIO & J. J. COSTANZI. 1977. Transplant. Proc. **9**: 1141.
13. SCHAFER, L. A., A. L. GOLDSTEIN, J. U. GUTTERMAN & E. M. HERSH. 1976. Ann. N.Y. Acad. Sci. **277**: 609.
14. SCHAFER, L. A., J. U. GUTTERMAN, E. M. HERSH, G. M. MAVLIGIT & A. L. GOLDSTEIN. 1977. *In* Progress in Cancer Research and Therapy. M. A. Chirigos, Ed. Vol. **2**: 329. Raven press. New York.
15. SCHAFER, L. A., J. U. GUTTERMAN, E. M. HERSH, G. M. MAVLIGIT, K. DANDRIDGE, G. H. COHEN & A. L. GOLDSTEIN. 1976. Cancer immunol. Immunotherap. **1**: 259.
16. LAVASTIVA, M. T., J. L. ROSSIO, A. L. GOLDSTEIN & J. D. DANIEL. Submitted for publication.
17. KLEIN, J. J., A. L. GOLDSTEIN & A. WHITE. 1965. Proc. Nat. Acad. Sci. USA **53**: 812.
18. KLEIN, J. J., A. L. GOLDSTEIN & A. WHITE. 1966. Ann. N.Y. Acad. Sci. **135**: 485.
19. GOLDSTEIN, A. L., F. D. SLATER & A. WHITE. 1966. Proc. Nat. Acad. Sci. USA **56**: 1010.
20. GOLDSTEIN, A. L., A. GUHA, M. M. ZATZ, M. A. HARDY & A. WHITE. 1972. Proc. Nat. Acad. Sci. USA **69**: 1800.
21. LOW, T. L. K. & A. L. GOLDSTEIN. 1978. *In* The Year in Hematology. R. Silber, J. LoBue, and A. S. Gordon, Eds. : 281–319. Plenum Pub. New York.
22. AHMED, A., A. H. SMITH, D. M. WONG, G. B. THURMAN, A. L. GOLDSTEIN & K. W. SELL. 1978. Cancer Treatment Eeports **62**: 1739.
23. SCHLESINGER, D. H. & G. GOLDSTEIN. 1975. Cell **5**: 361.
24. BACH, J.-F., M. DARDENNE & J.-M. PLEAU. 1977. NATURE **266**: 55.
25. WANG, S. S., J. D. KULESHA & D. P. WINTER. 1978. J. Am. Chem. Soc. **101**: 253.
26. FREIRE, M., O. CRIVELLARO, C. ISAACS & B. L. HORECKER. 1979. Proc. Nat. Acad. Sci. USA **75**: 6007.
27. OLSON, M. O. J., I. L. GOLDKNOPF, K. A. GUETZOW, G. T. JAMES, T. C. HAWKINS, C. J. MAYS-ROTHBERG & H. BUSCH. 1976. J. Biol. Chem. **251**: 5901–5903.
28. WATSON, D. C., W. BEATRIX LEVY & G. H. DIXON. 1978. Nature **276**: 196.
29. PAZMINO, N. H., J. N. IHLE, R. N. MCEWAN, T. L. K. LOW & A. L. GOLDSTEIN. Cancer Treatment Report. In press.
30. BACH, J.-F. & M. DARDENNE. 1971. Transplant. Proc. **4**: 345.
31. LAW, L. W., A. L. GOLDSTEIN & A. WHITE. 1968. Nature **219**: 1391.
32. HARDY, M. A., J. QUINT, A. L. GOLDSTEIN, D. STATE & A. WHITE. 1968. Proc. Nat. Acad. Sci. **61**: 875.
33. HARDY, M. A., J. QUINT, A. L. GOLDSTEIN, D. STATE, A. WHITE & J. R. BATTISTO. 1969. Proc. Soc. Exp. Biol. Med. **130**: 214.
34. GOLDSTEIN, A. L., S. BANEWEE, G. L. SCHNEEBELI, T. F. DOUGHERTY & A. WHITE. 1970. Radiation Res. **41**: 579.
35. GOLDSTEIN, A. L., Y. ASANUMA, J. R. BATITSTO, M. A. HARDY, J. QUINT & A. WHITE. 1970. J. Immunol. **104**: 359.
36. ASANUMA, Y., A. L. GOLDSTEIN & A. WHITE. 1970. Endocrinology **86**: 600.
37. BACH, J.-F., M. DARDENNE, A. L. GOLDSTEIN, A. GUHA & A. WHITE. 1970. Proc. Nat. Acad. Sci. USA **68**: 2734.
38. GOLDSTEIN, A. L., A. GUHA, M. M. ZATZ, M. A. HARDY & A. WHITE, 1972. Proc. Nat. Acad. Sci. USA **69**: 1800.
39. GOLDSTEIN, A. L., D. W. WARA, A. J. AMMANN, H. SAKAI, N. S. HARRIS, G. B. THURMAN. J. A. HOOPER, G. H. CQHEN, A. S. GOLDMAN, A. S. COSTANZI & M. C. MCDANIEL. Transplant Proc. **7(1)**: 681–686.

40. GOLDSTEIN, A. L., G. H. COHEN, J. L. ROSSIO, G. B. THURMAN, C. N. BROWN & J. T. ULRICH. 1976. Medical Clinics of North America. **60(3):** 591–606.
41. COSTANZI, J. J., N. HARRIS & A. L. GOLDSTEIN. 1978. *In* Immune Modulation and Control of Neoplasia by Adjuvant Therapy. M. A. Chirigos, Ed. : 373–380. Raven Press. New York.
42. COHEN, M. H., P. B. CHRETIEN, D. C. IHDE, *et al.* 1979. JAMA **241:** 1813.
43. LIPSON, D. S., P. B. CHRETIEN, R. MAKUCH, *et al.* 1979. Cancer **43:** 863.
44. GOLDSTEIN, A. L. 1978. Antibiot. Chemotherap. **24:** 47.
45. CHRETIEN, P. B., S. D. LIPSON, R. MAKUCH, *et al.* 1978. Cancer Treatment Reports **62:** 1787.
46. PERKINS, E. H. 1971. J. Reticuloendothel. SOC. **9:** 642.
47. GOLDSTEIN, A. L., G. B. THURMAN, T. L. K. LOW, J. L. ROSSIO & G. E. TRIVERS. 1978. J. Reticuloendothel. Soc. **23(4):** 253–266.
48. COSTANZI, J. J., R. G. GAGLUNO, F. DELANEY, N. HARRIS, G. B. THURMAN, H. SAKAI, A. L. GOLDSTEIN, D. LOUKAS, G. H. COEHN & P. D. THOMPSON. 1977. Cancer **40:** 14.
49. GOLDSTEIN, A. L., S. BANERJEE & A. WHITE. 1967. Proc. Nat. Acad. Sci. USA **57:** 821.
50. GOLDSTEIN, A. L., G. B. THURMAN, G. H. COHEN & J. A. HOOPER. 1975. *In* The Biological Activity of Thymic Hormones. D. W. Van Bekkum, Ed. : 145. Kooyker Scientific Publishers. Rotterdam.
51. MARSHALL, G. D., JR., T. L. K. LOW, G. B. THURMAN & A. L. GOLDSTEIN. 1978. *In* Human Lymphocyte Differentiation: Its Application to Cancer, INSERM Symposium 8. B. Serrou and C. Rosenfeld, Eds. : 271–277. Elsevier North Holland Press. Amsterdam.
52. PAZMINO, N. H., J. N. IHLE, R. N. MCEWAN & A. L. GOLDSTEIN. 1978. Cancer Treatment Reports **62:** 1749.
53. ZISBLATT, M., A. L. GOLDSTEIN, F. LIUY & A. WHITE. 1970. Proc. Nat. Acad. Sci. USA **66:** 1170.
54. HARDY, M. A., M. ZISBLAIT, N. LEVINE, A. L. GOLDSTEIN, F. LILLY & A. WHITE. 1971. Transplant. Proc. **3:** 926.
55. GOLDSTEIN, A. L., A. GUHA, M. L. HOW & A. WHITE. 1971. J. Immunol. **106:** 773.
56. ZATZ, M., A. WHITE, R. S. SCHULOF & A. L. GOLDSTEIN. 1975. Ann. N. Y. Acad. Sci. **249:** 499.
57. THURMAN, G. B., B. B. SILVER, J. A. HOOPER, B. C. GIOVANELLA & A. L. GOLDSTEIN. 1974. Proc. First Int. Workshop on Nude Mice. J. Rygaard and C. O. Povlsen, Eds. : 105. Gustav Fisher Verlag. Stuttgart.
58. THURMAN, G. B., A. P. STEINBERG, A. AHMED, D. M. STRONG, M. E. GERSHWIN & A. L. GOLDSTEIN. 1975. Transplant. Proc. **7:** 299.
59. DAUPHINEE, M. J., N. TALAL, A. L. GOLDSTEIN & A. WHITE. 1974. Proc. Nat. Acad. Sci. USA **71:** 2637.
60. GERSHWIN, M. E., A. AHMED, A. D. STEINBERG, G. B. THURMAN & A. L. GOLDSTEIN. 1974. J. Immunol. **11:** 1068.
61. TALAL, N., M. DAUPHINEE, R. PHILARISETTY & R. GOLDBLUM. 1975. Ann. N. Y. Acad. Sci. **249:** 438.
62. STRAUSSER, H. R., L. A. BOBER, R. A. BUSCI, J. A. SCHIUCOCK & A. L. GOLDSTEIN. 1971. Exp. Gerontol. **6:** 373–378.
63. SCHEINBERG, M. A., A. L. GOLDSTEIN & E. S. CATHCART. 1976. J. Immunol. **11:** 156.
64. FORGER, J. M. & J. CERNY. 1976. Cancer Res. **36:** 2048.
65. SCHEID, M. P., M. K. HOFFMAN, K. KOMURO, H. HAMMERLING, E. A. BOYSE, G. H. COHEN, J. A. HOOPER, R. S. SCHULOF & A. L. GOLDSTEIN. 1975. J. Exp. Med. **138:** 1027.
66. COHEN, G. H. & A. L. GOLDSTEIN. 1975. *In* The Biological Activity of Thymic Hormones. D. W. Van Bekkum, Ed. : 257. Kooyker Scientific Publishers. Rotterdam.

67. St. Pierre, R. L., S. D. Waksal, A. D. Barker, G. Cohen & A. Goldstein. 1975. Proc. Ninth Leucocyte Culture Conference, Williamsburg, Va. Abstract.
68. Barker, A. D. & V. S. More. 1977. *In* Control of Neoplasia by Modulation of the Immune System. M. A. Chirigos, Ed. : 289. Raven Press. New York.
69. Cercek, L., P. Milenkovic, B. Cercek & L. G. Lajtha. 1975. Immunology **29:** 885.
70. Miller, H. C., S. K. Schmiege & A. Rule. 1973. J. Immunol. **III:** 1005.
71. Goldstein, A. L., G. D. Marshall & J. L. Rossro. 1978. *In* Immunotherapy of Human Cancer. M. D. Anderson Hospital and Tumor Institute. : 173–179. Raven Press. New York.
72. Lowenberg, G., H. T. M. Nieuwerkerk & D. W. Van Bekkum. Personal communication.
73. Chirigos, M. A. Personal communication.
74. Asherson, G. L., M. Sembala, B. Mayhew & A. L. Goldstein. 1976. Eur. J. Immunol. **6:** 699.
75. Dabrowski, M. P. & A. L. Goldstein. 1976. Immunol. Commun. **5:** 695.
76. Naylor, P. H., H. Sheppard, G. B. Thurman & A. L. Goldstein. 1976. Biochem. Biophys. Res. Commun. **73:** 843.
77. Ahmed, A., A. H. Smith, D. M. Wong, G. B. Thurman, A. L. Goldstein & K. W. Sell. 1978. Cancer Treatment Reports **62:** 1739.
78. Ahmed, A., A. H. Smith, K. W. Sell, M. E. Gershwin, A. D. Steinberg, G. B. Thurman & A. L. Goldstein. 1977. Immunology **33:** 757.
79. Kenady, D. E., P. B. Chretien, C. Potvin, R. M. Simon, J. C. Alexander & A. L. Goldstein. 1977. Cancer **39:** 642–652.
80. Sakai, H., J. J. Costanzi, D. F. Louus, R. G. Gagliano, S. E. Ritzmann & A. L. Goldstein. 1975. Cancer **36:** 974-976.
81. Scheinberg, M. A., N. R. Blacklow, A, L. Goldstein, T. A. Parrino, F. B. Rose & E. S. Cathcart. 1976. New Engl. J. of Med. **294:** 1208–1211.
82. Moutsopoulos, H., K. H. Fye, S. Sawada, M. J. Becker, A. L. Goldstein & N. Talal. 1976. Clin. Exp. Irnmunol. **26:** 563–573.
83. Scheinberg, M. A., E. S. Cathcart & A. L. Goldstein. 1975. Lancet **1:** 424–429.
84. Rovensky, J., A. L. Goldstein, P. J. L. Holt, J. Pekarek & T. Mistina. 1977. Casopis Lekaru Ceskych **116:** 1063–1064.
85. Horowitz, S., W. Borcherding, A. V. Moorthy, R. Chesney, H. Schulte-Wisserman, R. Hong & A. L. Goldstein. 1977. Science **197:** 999–1001.
86. Horowitz, S. D. & A. L. Goldstein. 1978. Clin. Immunol. Immunotherap. **9:** 408–418.
87. Kaplan, J. & W. D. Peterson, Jr. 1978. Clin. Immunol. Immunopathol. **9:** 436.
88. Pazmino, N. H., R. N. McEwan & J. N. Ihle. 1977. J. Immunol. **119:** 494.
89. Goldstein, A. L. & J. L. Rossro. 1978. Comprehensive Therap. **4:** 49–57.
90. Astaldi, A. G. C. B. Astaldi, P. Wijermans, M. Groenewoud, P. Th.A. Schellekens & V. P. Eijsvoogel. 1978. J. Immunol. **119:** 1106.